The Book on 3D Printing

The Book on 3D Printing

by Isaac Budmen and Anthony Rotolo

DIGITAL FILE

SLICING AND GCODE OUTPUT

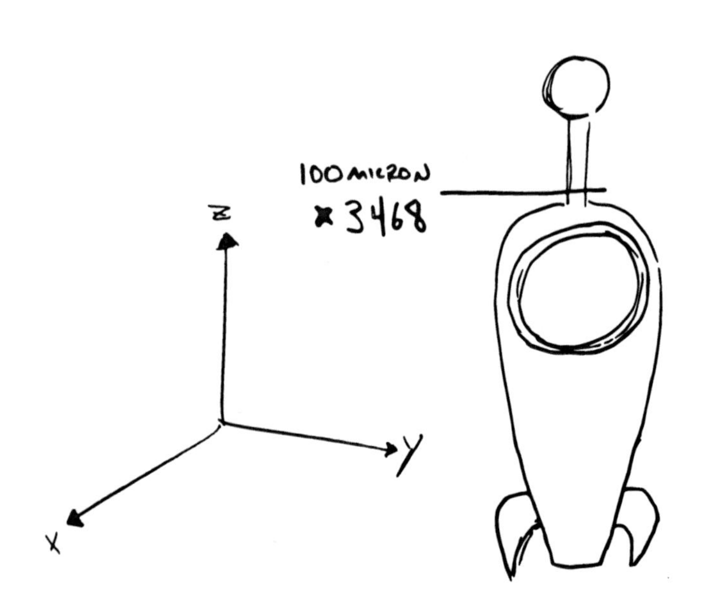

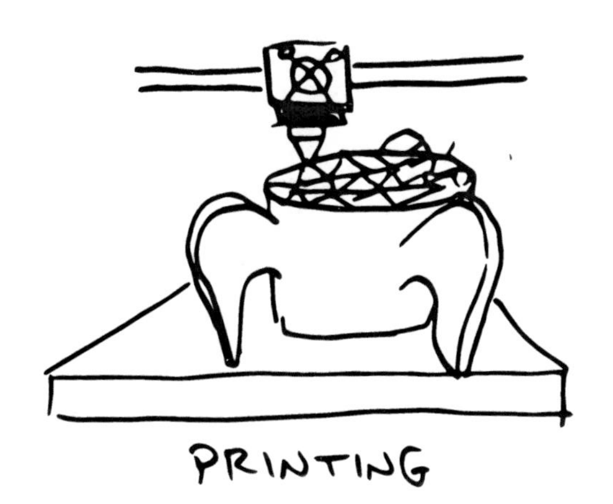

PRINTING

ENJOY
YOUR THING

ISBN-13: 978-1-489-52944-2
ISBN-10: 1489529446

Edited by the fearless and wonderful Diane Stirling

Illustrated by Isaac Budmen

Contents

To those we have pleased most of the time

A New Industrial Revolution

"Tea...Earl Grey...hot."

This command, famously given by Captain Jean-Luc Picard to the computer of the U.S.S. Enterprise, is much more than a drink order. Picard is speaking to a device called a "replicator," which is equipped with a database of recipes from across the galaxy. Responding to the Captain's voice command, the replicator identifies the correct design file for a cup of Earl Grey tea. It sets the custom parameter of temperature to "hot," per the Captain's request, and then produces the beverage on demand–cup and all.

Because the replicator works at the molecular level–translating the atoms of real life into the bits and bytes of digital information, and then back again–it can store the digital designs for everything from food and clothing to complex scientific equipment.

Star Trek's replicator, while a work of fiction, is perhaps the single best way of describing the goal of 3D printing. Those who are working in this exciting field have been chasing after Captain Picard's teacup for decades, believing that one day the dream of having "anything on-demand" will be a reality. That day is still far in our future, but advancements in material science, electronics and computing have made it possible to make some objects on demand. This work has given rise to machines that seem a lot more at home on the Enterprise than on your kitchen table.

3D printers, or machines that can fabricate physical objects, are surprisingly similar to Captain Picard's replicator. Just like their 24th Century cousins, a 3D printer is capable of reading a digital design file and following the necessary instructions to produce the item. Unlike the replicator of the future, which manipulates molecules to build an object, a 3D printer shapes physical materials like plastic, metal, wood or ceramic to design specifications. This means that today we can use a 3D printer to make a teacup, a saucer and even a spoon, but no machine can fabricate the tea from its basic molecular elements, and certainly not all at the same time.

Even if they can't yet make a cup of tea or a working electronic device, 3D printers can make an incredible array of things. The technology behind 3D printing, called *additive manufacturing,* is already in use to make some of the products you buy at the store. Recent advancements have allowed these same processes to be scaled down for personal use, giving artists, inventors and hobbyists the ability to produce the same kinds of objects inexpensively at work or home. Some even predict that 3D printing has the potential to spark a "new industrial revolution" of sorts, where individuals as well as corporations can manufacture goods.

In his fourth State of the Union address, President Barack Obama called for a rebirth of American manufacturing, pointing to this exciting but largely unknown new technology. The President described 3D Printing as having "the potential to revolutionize the way we make almost everything." This high profile mention ignited curiosity and speculation about 3D printing as many Americans and listeners worldwide heard about it for the first time.

President Obama helped introduce the idea of 3D printing to a mainstream audience, sparking the same curiosity that has captivated the tech industry for some time. But amidst the growing interest, and more than a few predictions of a 3D-printed future like something from a Star Trek episode, the details about this technology remain a mystery to many.

3D Printing sounds a lot more like science fiction than a modern marvel, but it is indeed a very real process that is already changing the way products are designed and manufactured. The purpose of this book is to provide a comprehensive look at 3D printing methods, machines and related products for those interested in wading into the early stages of this promising but experimental technology.

In the chapters that follow, we take a detailed look at Fused Filament Fabrication (FFF), the technical name for the plastic-based process most commonly known as "3D Printing," as well as some other methods of producing on-demand objects.

We wrote this book to include essential information for understanding 3D Printing, whether you are considering a 3D printer of your own, getting better acquainted with your machine or just curiously following the trends online.

The first part of the book serves as an introduction to the history and evolution of this technology and the current state of the industry. We will separate reality from hype to explain just what 3D printing is, what it isn't, and what's really possible to do with a 3D printer.

Once we've covered the basics, we will take a look at the most well-known machines and products currently available. This will help you decide which printer is right for you, no matter your work or interest area. Our look at machines currently on the market will also include some successful *crowdfunded* machines, or concept printers that have received financial backing from online supporters. These potential products offer a look into some new possibilities for 3D Printing.

No matter your printer type or model, a large part of 3D printing happens outside the machine. The process of designing and manipulating your own digital objects is where you take control of the things you will make. To make sure you have all the right tools at your disposal, we investigate the most popular 3D modeling software, as well as the array of software settings that can help you get the best objects from your printer. We will also take a look at the growing number of communities and resources emerging to help hobbyists and pros exchange ideas and support information.

If you picked up this book to learn about 3D printing as a concept, but not necessarily to learn about your own machine, we recommend skipping the second part of this book for now (just jump right to the closing chapter). We wrote the chapters in Part 2 for those of you who want to understand the inner workings of an FFF 3D Printer.

Our experience with homebrewed and commercial machines has taught us that successful 3D printing involves developing a relationship with your printer and its mechanics. To help you become an expert operator of your machine, we will explain the common components of all 3D printers in detail, as well as the various options and possible upgrades currently available on the consumer market. Part II: Inside a 3D Printer should also come in handy if and when you encounter performance issues with your printer.

Part I: Understanding 3D Printing

1. Introduction to 3D Printing

3D Printing is a method of producing a physical object from a digital file. It is achieved through a process of precisely forming a raw material into the desired, solid shape. The term "3D printing" itself may be a bit misleading, conjuring images of an inkjet printer that comes with special glasses, but this is because the concepts behind 3D printing are not all that different from the desktop publishing devices found in our homes and offices.

Like many other types of printers, 3D printers are computer numerical controlled (CNC) machines that follow a set of instructions to trace a specified design. The biggest difference between 2D and 3D printing is the addition of a Z-axis–a third dimension–that allows either the print head or platform to raise and lower, giving the machine the ability to build the height or depth of an object in addition to its length and width.

Interest in 3D printing has been steadily growing in recent years following the release of desktop machines priced within reach of some consumers. With home and small business use increasing, references to the technology are showing up everywhere from news reports to tech blogs. CBS's *The Big Bang Theory* even featured a 3D printer in one episode, which was depicted as an expensive and not particularly useful technology that is nonetheless amazing.

The September 2012 issue of Wired Magazine featured a cover story on 3D Printing and its potential to change the world. This and other reports point to a near future filled with objects, foods and even synthetic body parts printing on demand and accessible to all. However, the current reality of 3D printing is quite different from this fantastical, although entirely possible future.

Today's 3D printing, like the early days of the personal computer or the mobile phone, is a bit less glamorous. Early adopters and tech enthusiasts tinker away with kits and new consumer-focused machines that produce mostly small, plastic objects from the imagination of hobbyist designers. These tiny creations–the toys, figurines and such that Slate Magazine called "useless plastic trinkets"–have become commonly associated with desktop 3D printing[1].

Despite the technical challenge and material limitations of current desktop fabrication, watching a 3D printer make an object for the first time is a wondrous experience. Witnessing a real, tangible object emerge before your eyes from layers of liquid plastic or a pool of resin feels like meeting the future head on. Perhaps this is what has captured the imagination of so many who have worked to develop and advance this technology, and the proud owners of the first commercially available machines.

Types of 3D Printing

What we now call "3D Printing" is one type of computer-controlled manufacturing process used to create smaller-scale objects from a digital design file. The most common

1 Oremus, Will . "The Steve Jobs of Useless Plastic Trinkets." Slate, September 21, 2012. http://hive.slate.com/hive/made-america-how-reinvent-american-manufacturing/article/the-steve-jobs-of-useless-plastic-trinkets.

form of 3D printing is called *additive manufacturing.* Additive manufacturing machines usually extrude, or squirt, a plastic material to "print" objects. This is the method employed by nearly all consumer-oriented 3D printers and a number of industrial machines.

The most common form of additive manufacturing used in 3D printing is known as *Fused Filament Fabrication*, or FFF. This process, sometimes referred to as *Fused Deposition Modeling* (FDM), involves the use of a thin, plastic-based filament which is heated and extruded through a print head. The technique is not unlike the method of printing ink on paper, except that FFF printers repeatedly draw one layer on top of the next until a three-dimensional object is formed.

Fused filament fabrication is the technology at the heart of nearly all consumer 3D printers and is the focus of this book. FFF has been advanced by a community of scientists, engineers and developers working on largely open-source projects. Even as proprietary 3D printers enter the consumer market, such as the Makerbot Replicator 2 and the Cube, this technology continues to play a central role in the most widely available machines. We will explore FFF technology in greater detail in the next chapter.

In addition to FFF, several other types of 3D printing are in use or under continued development. These include *stereolithography* (SLA) and *selective laser sintering* (SLS), which are additive manufacturing processes that form objects from a pool of curable material instead of extruding plastic as in FFF/FDM machines. SLA printing is achieved by building object layers from a liquid polymer resin using ultraviolet light to cure or harden the material. The process was first invented and patented in the 1980s and has seen some adoption in industrial machines.

SLS technology has also been around since the 1980s, developed at the University of Texas at Austin as a DARPA-funded project. SLS is an additive process that uses a laser to form three-dimensional shapes from a powder material. SLS machines are capable of creating objects from a range of materials, including metal, ceramic and plastic. Both SLA and SLS machines are usually suited for large-scale manufacturing settings and can be priced at well over $100,000.

Much of the progress made over the last decade toward home or "desktop" 3D printers can be attributed to a few important factors. First is the expiration of a number of patents on processes and technologies that, until recently, have seen limited development by small groups behind closed doors. With fused filament fabrication and similar methods available for open-source collaboration, the evolution of additive manufacturing accelerated dramatically.

The other driving force behind 3D printing has been the progress made by the RepRap project. RepRap, which stands for *Replicating Rapid Prototyper*, is an open-source community that was founded in 2005 with the goal of creating a machine that can replicate most or all of its own components. Focused on *Fused Filament Technology* (a variant of FDM, which is a trademarked name), RepRap has produced a number of important machine designs that serve as milestones in the evolution of 3D printing.

All work done by RepRap developers and designers is released under a free, public license. This means that the results of RepRap research and development are available for anyone to use or expand upon. This has led to tremendous interest and rapid advancement in just a few years, which is especially impressive when compared to technologies developed in closed or proprietary environments.

Much like the computer clubs of the 1970s and 80s, where some of the industry giants got their start, the RepRap community has had several members go on to develop and sell consumer 3D printers. Through online collaboration and resource sharing, the distributed 3D printing community has been able to thrive. The nature of 3D printers and their ability to self-replicate have made it easy for developers to share new designs that can be printed (instead of manufactured and shipped) anywhere. This means a talented industrial designer in New Zealand can invent an extruder upgrade that can be replicated and used by anyone with a comparable machine, just by sharing the digital design files.

Around the same time the RepRap project was beginning, another technology was introduced that would make developing the electronic components of a 3D printer much less complicated. The *Arduino,* which is an open-source micrcontroller board, has become the engine in nearly all consumer 3D printers.

With Arduino, the 3D printing community was able to prototype and develop working, computer-controlled machines without the level of electronics expertise previously required for projects of this scale. Arduino opened the door for everyday developers to control an endless array of sensors for the first time, which has led to a wave of new consumer electronics developed by small companies and inventors.

Many current 3D printer models are based on "open-source" designs, whether the FFF technology and machine configurations or the Arduino-based electronics used to control the device. In fact, some of the most successful printer models thus far are derived directly from the work of the RepRap movement. This includes the Replicator series from Makerbot Industries, although the company has begun patenting some aspects of its latest models.

According to the Open Source Hardware Association (OSHWA), the term "Open Source Hardware" refers to "machines, devices, or other physical things whose design has been released to the public in such a way that anyone can make, modify, distribute, and use those things."[2] The idea behind open-source hardware is that by having a community of contributors working freely on a technology, it can be developed more quickly and effectively than a closed, patented technology that is developed by one person or company. Much like the open-source software movement, which developed the Linux operating system, among other achievements, open-source hardware is a controversial but exciting approach to product development.

The movement associated with these ideals is sometimes called the "Maker Movement," but this term is not limited to 3D printing. Makers from all over the world are working on electronics and other devices that are the work of hundreds (even thousands) of programmers, designers and engineers contributing freely online. New business are emerging from this work, including some niche electronics that have enjoyed rapid growth due to the passionate, open-source communities around them.

Uses for 3D Printing

3D Printing, or additive rapid prototyping, has been used in industrial settings for some time now. Some of the most common applications are currently found in the automotive industry, where the ability to quickly build concept parts and models is a powerful asset. The

2 Open Source Hardware Association, "Open Source Hardware Definition." Accessed May, 2013. http://www.oshwa.org.

same is true for a range of consumer products that can be designed or even manufactured for sale using additive machines.

3D printing is also finding its way into the aerospace industry, including some promising applications at NASA. Just as RepRap visionaries imagined a machine that could replicate essential supplies, or even itself, NASA imagines that one day 3D printing may allow for the fabrication of objects in space. This could be a game-changing technology for space exploration, eliminating the need to carry some objects into space.

Until that is possible, NASA is also taking advantage of 3D printing here on Earth. You are likely to spot a number of 3D printers around NASA space centers as teams from across the agency familiarize themselves with the new technology.

Home applications for 3D printing are a bit different from the prototyping and replicating for which they were invented. Hobbyists, artists and inventors will certainly find the ability to quickly and inexpensively produce custom objects to be a boon for their creativity, but the majority of 3D printer owners (an estimated 32,000+ by the end of 2012) report using their machines for a variety of purposes.

Among hobbyists, model making and artwork remain two of the most common uses for a 3D printer. Some creative home users have even designed replacement parts or upgrades for lost or broken items. This might include battery covers for the television remote or a missing button, handle or knob. Others design and create dollhouse furniture, figurines, molds and patterns, jewelry and mobile phone accessories.

Whatever the creation, the hobbyist 3D printing community has begun sharing these objects using online forums. There are a number of trinkets and fun objects that have spread through this small group of enthusiasts, like the Heart Gears (a moving set of gears shaped like a heart) or the Stretch Bracelet (an elastic wristband made from plastic printed one layer thick). Both of these items were designed and shared online by Emmett Lalish, an engineer and 3D printing hobbyist from Seattle. These items have been printed by 3D printer owners worldwide and are two of the most common items used to demonstrate the capabilities of a desktop 3D printer.

3D Printing Market

The term "3D printing" has become popular in the media and among tech communities in recent years. Word has spread through pop culture references, celebrity hobbyists like Jay Leno and even a music video by will.i.am featuring a 3D printer. Still, most people have not yet experienced a 3D printer in person, and far fewer have had the chance to use one.

In a recent survey, conducted by the Peer Production Foundation researchers were able to find only 350 respondents who owned a 3D printer.[3] This study, along with others focused on the additive manufacturing industry, has estimated the total number of consumer 3D printers sold to date is 32,000-70,000 units.[4] These numbers represent a very small market for a technology that remains in an early stage of development.

Many have compared the current state of the 3D printing market to the personal computer market in the

3 Moilanen , Jarkko, and Tere Vadén. P2P Foundation, "Manufacturing in motion: first survey on 3D printing community." Last modified May 31, 2012. http://surveys.peerproduction.net/2012/05/manufacturing-in-motion/.

4 Wohlers Associates, *Annual Worldwide Progress Report*. 2013. "Additive Manufacturing and 3D Printing State of the Industry." http://wohlersassociates.com/2013report.htm

late 1970s and early 1980s. At that time, most personal computers were homebrewed machines or assembly kits purchased by hobbyists for the sake of learning about the new technology. These early PCs were a rather expensive hobby for most at the time, especially because there was no commercial software available to run on the computer. This was the time of computer clubs and user groups, when the machine we would someday know as the personal computer was still a collection of parts and wires in Steve Jobs' garage.

Indeed the same scenario describes 3D printing today. At this time, much of the activity and sales around desktop 3D printing are concentrated on a small group of enthusiasts and tinkerers who have purchased or assembled a machine to explore this new technology. They meet in online communities and at local makerspaces to share their experiences, hacks and upgrades, and many continue to further develop the technology in their home workshops and certainly a few garages.

Perhaps the most important similarity between personal computers and 3D printers is the level of interest and excitement that surrounded both technologies in their early stages. For personal computers, the interest continued to grow as the industry finally reached an ownership rate of more than one PC per household in the United States by 2010.[5] It is unclear whether 3D printers will follow a similar path, or if it will take nearly 30 years to see one in every home. Whenever that day arrives, it is safe to say that the household 3D printer of the future will be quite different from the early machines of today.

5 Economist Intelligence Unit, December 2009

2. Fused Filament Fabrication

Although the term may seem rather obtuse at first, Fused Filament Fabrication is a straightforward process that works just as its name suggests.

Fused: joined together forming a single object
Filament: a threadlike material
Fabrication: the act of making a physical object according to predefined specifications

If you've ever done a project involving hot glue, then you know that a hot glue gun works by melting a solid stick of glue and then squeezing the liquid glue through a nozzle. This allows the user to apply the glue rather precisely. Likewise, you can think of the FFF process as a glorified hot glue gun. In an FFF-based 3D printer, an extruder pulls in plastic filament (much like a hot glue stick) and forces it through a very hot metal nozzle in a specific pattern.

Invented in 1989 by S. Scott Crump as a means of rapid prototyping, the process was initially known by the name *Fused Deposition Modeling* (FDM), which Crump trademarked as his company began manufacturing industrial FDM machines. This is the most popular type of 3D printing on the consumer market, and is the most commonly referenced additive manufacturing process.

3D printing materials are most often found as spools

of plastic *filament,* the most common among them are two types known as *ABS* and *PLA*. These materials each offer properties that make them useful for making a variety of objects, and each have been used in additive manufacturing for some time.

There are also a number of experimental filaments emerging as interest and activity around 3D printing grows. All of these filament types are available online through a number of distributors. Consumer printer manufacturers may also sell their own brand of filament, but generally it is not necessary to stick to manufacturer materials.

Acrylonitrile butadiene styrene (ABS), is a popular plastic that has been used to make all sorts of home goods and toys, including LEGO building blocks. Its durability and resistance to impact make it a versatile option. ABS is petroleum-based and gives off a toxic fume when heated, although this downside can be mitigated with proper ventilation to the machine and surrounding area.

ABS is often used for injection molding, which is the process by which many plastic products are currently manufactured. This is because ABS plastic shrinks slightly when it cools, making it easy to remove from the mold. However, this characteristic is not ideal for 3D printing, making ABS a more challenging material to use with FFF technology.

When used for 3D printing, ABS shrinkage causes the object to curl off of the build platform, leaving you with a warped object. To combat this, several methods have been invented by hobbyists and manufacturers working with the material. The most effective method of preventing shrinkage is the use of a heated build platform or build chamber inside the 3D printer. This can be imagined much like building your object on a hotplate or inside of a low-temperature oven. The heat from the build surface helps the ABS cool evenly with less warping.

Polylactic acid (PLA) has recently emerged as a favorite material for 3D printing. PLA is a biodegradable plastic made from starch derived from corn and other crops. PLA melts at slightly lower temperatures than ABS, requiring a lower temperature setting on a 3D printer. However, the biggest difference between PLA and ABS plastic is that PLA can be extruded at room temperature with little to no warping.

PLA is a very strong and rigid plastic, making it superior to the softer and more flexible ABS for printing some parts and objects that must withstand more rugged situations. However, printing moving objects or interconnecting parts with PLA requires very accurate tolerances as its strength makes it less forgiving.

Polyvinyl acetate (PVA) is also available as a printable filament. PVA is a flexible polymer material that has long been used to make many common glues and adhesives, including the popular Elmer's Glue and a number of wood glue products. It is more sensitive to temperature and humidity than other filaments, but it offers the unique ability to dissolve in water at room temperature.

Water solubility gives the machine operator the ability to build complex objects that require support structures to print correctly. By printing support material using PVA, the supports can be easily dissolved overnight in water once the print has finished.

Support material can also be printed using only one type of filament (ABS or PLA, for example), but needs to be removed by hand and can require filing in hard-to-reach places.

PVA is most often used in conjunction with ABS or another polymer. This is achieved by adding an additional print head to the machine, much like a color inkjet printer uses several print heads and cartridges. PVA must be used at exactly 190°C. Beyond that temperature, the material begins to rapidly degrade and can cause jams. We will explore support materials and additional extruders in Part II of the book in greater depth.

Experimental Materials

The focus on ABS and PLA printers has allowed for significant growth and creativity within the market, but it has also limited the number of real-world applications for 3D-printed things. To go beyond the capabilities of plastic, the community has done considerable research and development of flexible, fibrous and even conductive filament materials. The introduction of new options to the palette of printable materials greatly expands the versatility and future potential of 3D printers.

Wood fiber filament is capable of forming objects that look, smell and feel like wood. These objects can also be handled like wood, allowing them to be cut, sanded and painted in the same way. This filament is made of a PLA-like polymer and recycled wood fiber. It should be used between 175°C and 250°C.

Depending on the print temperature used, the filament will take on various shades of brown. This means that by varying the printing temperature throughout the print, you can achieve a variety of striping and shading effects. The rule of thumb is the lower the temperature, the lighter the finish.

Nylon is a durable, flexible and chemically resistant material that can be used to make everything from t-shirts to pipe fittings. It's no surprise, then, that the ability to print with nylon has awakened the creativity of 3D printing enthusiasts. Everything from wearable belts to orthopedic inserts are possible with nylon, and the filament can be colored using clothing dye found in craft stores.

Nylon is optimally printed between 245°- 265°C, and on a cellulose dense surface so that the material can bond to the platform. Very dense cellulose materials like Garolite will create strong bonds with nylon printed objects. This process is used for precision part-making. Large areas of contact between the nylon object and a Garolite surface can actually form a nearly permanent bond between them. On the opposite end of the spectrum, poplar wood and various tapes containing cellulose will also allow the nylon to bond, but without such a firm grip. However, these materials run the risk of the object coming loose from the build plate during printing.

Polycarbonate is a strong, shatter-resistant filament. Despite its reputation for toughness, a hard polycarbonate is surprisingly soft enough to be gripped by the drive wheel of a filament extruder. Polycarbonate filament is still in the early stages of experimentation and is usually sold with a disclaimer about the bleeding-edge nature of the material. The filament needs temperatures upwards of 265°C in order to print. This isn't possible for most desktop 3D printers without causing damage to the machine.

Conductive filament is still in the academic paper phase, with limited retail availability. Conductive filament is made from an ABS-type polymer with blends of carbon fiber and a conductive granular material. In theory, the material's properties can allow for printing of functioning electronics. Uses could range from anti-static applications to conduction of electric current. Resistance values of conductive filament have been reported at approximately 100 Ohms/cm. Like ABS, it is used at 200°C-220°C

3. Machines and Manufacturers

The consumer market for 3D printers remains small in relation to other consumer electronics. As of 2012, the total number of desktop 3D printers sold is estimated to be 70,000 worldwide. In comparison, more than 10 million 2D printers (desktop publishing) are sold each year.[6] Still, there are a number of 3D printer manufacturers with products available or pending release.

Companies who make consumer 3D printers range from kit makers at their kitchen tables to starups like Makerbot Industries, which has received significant venture capital to support its line of machines. Larger companies with experience in the industrial market have also released consumer-oriented machines, but it is unclear whether significant resources are being dedicated to the products in some cases.

The most common Desktop 3D printers can be broken into three categories. The first are the prefabricated machines that offer printing as soon as the product is unboxed. The second group are the assembly kits, which are popular among hobbyists who don't mind building their own printer. These machines are often available for a lower price and sometimes offer more choice or the latest innovations for those who are willing to experiment. Finally, there are the crowdfunded machines. These are primarily conceptual products that

6 Jamieson, Larry. Photizo Group, 2012.

have received investment from backers on Kickstarter and similar sites. We will take a look at crowdfunded 3D printers in the next chapter.

Pre-fabricated 3D Printers

Makerbot Industries, based in Brooklyn, New York, produces the "Replicator" line of printers that together make up the largest share of the consumer market. Its current flagship machine, called the Replicator 2, is a PLA-based machine with one of the largest build volumes available in a pre-assembled 3D printer. Some aspects of the product design are proprietary to Makerbot, which represents a departure from the company's open-source roots that made them an early favorite of the community.

Building off the contributions of the RepRap project, founders Bre Pettis, Adam Mayer, and Zach Smith recognized that sourcing obscure parts and expensive components had become a challenge for individuals interested in building a printer of their own. As a result, Makerbot was one of the first companies to begin packaging kits for hobbyist assembly, lowering the barrier to entry for 3D printing.

Makerbot officially entered the market in March 2009 with the introduction of its first desktop machine, the Cupcake CNC. Advertised as "an open, hackable robot," the Cupcake CNC incorporated many of the lessons learned from the RepRap project and included all the pieces necessary to build a 3D printer. However, unlike the RepRap project, Makerbot did not set the goal of producing a machine that could replicate itself. Instead, the goal was to manufacture a kit for hobbyists everywhere to hack and assemble.

The Cupcake CNC featured a 4x4x5-inch build platform with an optional heated bed. As a fully open-source machine, designs were available for anyone with the means to source the components and independently build or customize their own Cupcake CNC.

Following the maturation of the Cupcake CNC kit and open-source community, which had developed improvements, adjustments and redesigns, the company released the Thing-O-Matic in September 2010. The machine featured a chassis made of laser-cut plywood and a larger build volume than its predecessor. This was Makerbot's first model to offer a heated build platform upgrade making the machine capable of producing objects using either ABS or PLA plastic.

The Thing-O-Matic also featured an optional "automated build platform" upgrade. The feature added a small conveyor belt to the build platform that would slide a finished object off the bed. Unfortunately, the automated build platform was plagued with performance issues and was shelved when Makerbot discontinued the Thing-O-Matic in 2011. It is estimated that Makerbot produced about 850 Thing-O-Matic kits between 2009 and 2011.

Makerbot's first real departure from the original design of the Cupcake CNC came in January 2012 with the release of its third machine called the Replicator. It was the first pre-assembled printer from Makerbot. Although the design of the Replicator remained entirely open-source, Makerbot began assembling and shipping the product at its headquarters in Brooklyn. This step was partly due to the increased complexity of the machine compared to previous generations and was a welcome change for customers looking for a machine that arrived ready to print.

The Replicator featured a larger heated build platform (11.2 x 6.0 x 6.1 inches) than the Thing-O-Matic and had the option for dual extruders. Upgrading to dual print heads allowed operators to print two-color objects, or to print support structures with water-soluble PVA. Despite this capability, a single extruder using ABS or PLA remained the most popular configuration for Makerbot operators.

In mid-2012, just months after launch, cloned versions of the Replicator began appearing on the market. Based on the machine's open-source designs, these nearly identical printers were offered by a competitor supplier for well under Makerbot's retail price. The number of cloned Replicators sold or their impact on Makerbot's bottom line is not known, but it is clear that the experience prompted Makerbot to re-evaluate its business model. The company's next steps sparked debate around open-source business and questions of viability.

Later that year, Makerbot introduced the Replicator 2, the successor to the nine-month-old Replicator. The new machine once again increased the size of the build volume from previous models, eliminated the option for dual extruders and removed the heated bed. These changes meant the Replicator 2 was capable of working with only PLA plastic.

The Replicator 2 is aesthetically different from the models that preceded it. Its powder-coated steel chassis and glowing LED lights helped to attract the attention of gadgeteers and tech industry press. However, the Replicator 2 contained largely the same components and electronics as the previous model. The most significant change was not a new feature or the machine's sci-fi appearance. With the Replicator 2, Makerbot began to close the source on their product. In this case, the designs for the electronics and steel chassis of the Replicator 2 were patented and not released to the public.

The company's choice to move to a proprietary business model largely overshadowed the technical abilities of the Replicator 2 upon its release. Criticism over the move and what it meant for the open-source community, which remains an important portion of the 3D printing audience, forced Makerbot CEO Bre Pettis to defend the decision at the 2012 Open Hardware Summit in New York City. In a similar blog post later that month, Pettis denied claims that his company was benefiting from the work of others, saying, "we are not abandoning the RepRap community. In fact we believe everyone involved should be very proud of what we (the community) and Makerbot have accomplished."[7]

Along with the controversy around its business model came a rash of technical issues for Replicator 2 owners. The printer shipped with a plunger-based extruder, called MK7, which was notoriously plagued with jamming, stripping and clogging issues. Users quickly developed their own spring-based extruder solutions which were shared online and widely considered to be necessary for proper operation. Although the Replicator 2 is marketed as a user-friendly desktop 3D printer, this proved to be an optimistic description for those who had hoped for a reliable, consumer-friendly product.

Filling the gap between the Replicator and the Replicator 2, Makerbot released the Replicator 2X in late 2012. Described as an "experimental" 3D printer, the machine features an identical steel chassis to the Replicator 2, but the dual extrusion and heated build abilities of the original Replicator were once again added to the configuration. The 2X also featured an updated extruder that used a spring mechanism based on the solution designed by the open-source community to address flaws in the Replicator 2 extruder.

Makerbot Industries was acquired in the spring

7 Bre Pettis, "Let's try that again," *Makerbot Blog* (blog), September 24, 2012, http://www.makerbot.com/blog/2012/09/24/lets-try-that-again/.

of 2013 by Stratasys Ltd, maker of industrial additive manufacturing machines. Makerbot continues to operate as a subsidiary of Stratasys, however future plans for the brand and current products lines are not yet known.

Another notable manufacturer of pre-fabricated 3D printers is 3D Systems. Based in Rock Hill, South Carolina, 3D Systems has been producing industrial machines since the mid-1980s. The company is best known for its large-scale printers focused on stereolithography (SLS) technology, which was invented and patented by the company's founder, Chuck Hull.

3D Systems is quite different from Makerbot in its corporate structure and approach to consumer products. As a publicly traded company (NYSE: DDD), 3D Systems employs over 1000 people as of mid-2012, which is the result of several strategic acquisitions in recent years. Keeping true to its roots in proprietary business machines, 3D Systems launched its line of patented consumer 3D printers to address the growing hobbyist market.

Cube is marketed by 3D Systems as a "plug and play" 3D printer at an affordable price. Coming in at nearly $1000 less than the retail price of a Makerbot Replicator 2, the product was welcomed by bloggers and fans of 3D printing with hopes that new users might find Cube to be an approachable entry point to the technology. 3D Systems' history with business machines was also seen as an advantage in this area. Unlike Makerbot, which uses a rigid, kit-based design and futuristic aesthetic, the Cube is a truly prefabricated machine. Wrapped in a bright-colored, plastic chassis, the Cube looks a lot more like a familiar retail product than other 3D printers.

The feature set of the Cube is also somewhat different. As Makerbot has consistently increased the build volume of its printers, Cube has chosen to keep a smaller build platform (5.5 x 5.5 x 5.5 inches) in their desktop model.

This means that Cube printers are limited to making somewhat smaller objects than comparable machines. In addition, Cube includes a heated build plate, making the printer capable of working with either PLA or ABS filament.

In 2012, 3D Systems released a pro-class desktop printer called Cube X. This machine offers an optional second print head and was the first consumer 3D printer to offer three heads, if desired. This makes the Cube X a versatile machine, allowing for three-color prints, or the use of dissolvable support material. However, the Cube X is priced even higher than the Makerbot Replicator 2X for a dual-extruder model. With the addition of a third extruder, the Cube X becomes a far less affordable option for personal use.

Beyond minor variations in features and specs, the differentiator in the pre-fab 3D printer category may not be about the machines at all. There is a clear contrast in the way Makerbot and 3D Systems approach the community of hobbyists, enthusiasts and artists that surrounds each brand. Makerbot has developed a social networking site called *Thingiverse* where designers of 3D objects can share their creations. Cube owners, however, will once again find a retail-oriented approach. Cube's website hosts objects from Cube Designers as paid downloads. We will take a closer look at online communities in the final section of this book.

Kit-type 3D Printers

Bridging the gap between pre-fabricated and assembly kit 3D printers is Ultimaker, makers of a machine by the same name. Based in The Netherlands, Ultimaker began selling its first 3D printer in March 2011 and was greeted warmly by hobbyists who purchased nearly 3,000 first-run machines.

Erik de Bruijn, Martijn Elserman and Siert Wijnia, the team behind the Ultimaker, gained experience in 3D printing through their active involvement in the RepRap open-source community, working toward building a machine capable of replicating itself. The three applied their extensive knowledge with additional experience designing for 2D printers and laser cutters to create the Ultimaker.

Ultimaker is an assembly kit printer, meaning that the machine is sold as a kit containing all the parts needed for the owner to assemble it at home. Much like Swedish furniture-maker Ikea, Ultimaker is able to lower the price of its product without sacrificing quality by shipping the item "flat-packed" and leaving the assembly work to the customer. In the case of Ultimaker, the same printer is also sold pre-assembled for an additional fee, allowing for immediate printing right out of the box. This approach has helped Ultimaker appeal to a range of customers with a choice of an attractively priced kit option or the completed machine that is priced competitively against pre-fab machines from Makerbot and 3D Systems.

The features of the Ultimaker make this printer a very strong contender in the current market. It includes a large build volume (about 8x8x8 inches) that allows the machine to produce taller objects (about 512 cubic inches in volume) compared to the Replicator 2's build volume (approximately 396 cubic inches). Speed is also an advantage for the Ultimaker. It features a moving Bowden extruder (see Chapter 10) and very fast horizontal acceleration, making it capable of printing objects faster than other machines.

Lastly, the Ultimaker is capable of 50-micron print accuracy in all directions. A *micron* measures one milionth of a millimeter and it is the unit used by many 3D Printer manufacturers to describe the finest resolution capable with the machine. A common default resolution for a desktop machine might be 200 microns.[8]

Type A Machines is a relatively new player in the 3D printer market. Founded in mid-2012, Type A turned heads when its Series 1 printer received the "Best in Class" award from O'Reilly Media's "Make Magazine." The Series 1 is similar in design to other plywood assembly kit printers but is sold pre-assembled and ready to print. Despite the additional resources needed to produce pre-assembled machines, Type A offers the Series 1 for $1,400, which is a very aggressive price point when compared to comparable offerings from its competitors.

The Series 1, with its plywood chassis and utilitarian design, is certainly not the prettiest 3D printer on the market. Instead, Type A Machines' founder, Andrew Rutter, set out to create a reliable machine priced within reach of hobbyists looking for printing performance. This includes one of the largest print volumes available (9x9x9 inches) with print speed and quality that matches or surpasses more expensive printers.

8 Regardless of impressive specs like the one from Ultimaker, or the 100 micron setting on the Makrbot Replicator 2, it is important to note that high-resolution printing at 50 or even 100 microns is difficult to achieve with current models.

4. Crowdfunded 3D Printers

Crowdfunding sites like Kickstarter and Indiegogo have played a significant role in the 3D printing arena. These sites help individuals or companies with product ideas obtain financial support from everyday investors, or "backers," to produce the idea. Campaigns seeking as little as a few thousand or as much as several hundred thousand dollars have been funded through this method.

Crowdfunding achieved mainstream attention after a number of high-profile success stories, including the Pebble smart watch and a fan-supported movie based on the Veronica Mars television show raised millions of dollars on Kickstarter. While these results are not typical for a crowdfunding campaign, products and projects from technology to art and music continue to find a supportive audience in these online communities.

Crowdfunding works by offering aspiring product designers a platform for presenting their idea and demonstrating the ability to deliver the product. Potential backers peruse these profiles looking for product ideas they might purchase if the item were available. Backers are then able to commit funding for projects at a variety of supporter levels. Backers are not charged unless the project reaches a predetermined minimum level of funding, so there is no loss if the funding campaign does not succeed.

Backers on crowdfunding sites are motivated to

donate through rewards offered for their chosen level of funding. In the case of 3D printers and other electronics, the most common funding option is a pre-order price on the device itself. Backers of a successfully funded 3D printer are therefore the first to own the new machine. The risk here is that a funded Kickstarter campaign has no obligation to actually deliver the product, so a backer could be left empty handed.

The odds of being disappointed are high according to one study at the Wharton School of Business, which found that only 25% of successful Kickstarter campaigns in the technology and design category have delivered their products thus far. [9]

This low rate of return can be explained in several ways. In some cases, the reality of producing the product has proven more difficult than anticipated. This was the case for the Oculus Rift, a virtual reality headset from gaming startup Oculus VR. The team raised $2.4 million on Kickstarter but had anticipated building far fewer devices than the 7,500 units ordered by its backers. Oculus has not announced a delivery date for the Rift product, which was funded in September 2012. However, the company has given a release date of June 2013 for the Oculus Rift developer kit, which indicates that the retail version is not likely to appear until late 2013.

The Oculus Rift is not the only Kickstarter favorite to face production challenges. Gadgets like the Touch Time watch and the Galileo iPhone camera mount were big winners in the crowdfunding lottery, and each encountered significant delays. Even the Pebble smartwatch, which remains one of Kickstarter's most successful products, failed to ship a finished product to eager backers on schedule. Considering the more than

9 Mollick, Ethan R., The Dynamics of Crowdfunding: An Exploratory Study (June 26, 2013). Journal of Business Venturing, 2013. http://ssrn.com/abstract=2088298

$10 million raised by Pebble, and the wave of venture capital that followed, it is easy to imagine the challenge ahead for products that have raised far less.

The reality of crowdfunded ideas is that few are developed beyond anything but a good idea when funds are received. The individuals and teams that find success on Kickstarter are suddenly thrust into a whirlwind process of forming a company, hiring their first employees and figuring out how to scale their prototype (if one exists). For some, like Pebble, this includes setting up production in China or another "manufacturing country," as opportunities to make consumer electronics in the United States are limited and cost prohibitive. Many startups lack the experience or resources necessary to navigate these tasks, and some find themselves scrambling to learn on the job.

Despite the uncertainty of crowdfunding, the excitement around 3D printers has not slowed down on Kickstarter. Several potential products have been successfully funded, including interesting new applications for 3D printing technology and machines that promise top performance for an affordable price.

Once such example is Robo3D, which asked the Kickstarter crowd for $49,000 and raised just over $649,000 from more than 1,200 backers. The concept for the Robo3D was developed by three recent college graduates from Southern California to create a printer that would match or exceed the features of the Makerbot Replicator 2 at an incredible price point of just over $500.

At this time, Robo3D remains in "pre-production," according to an update sent to the group's backers. In the message, Robo3D states that their original goal of bringing a "nice, affordable 3D Printer" to the market would not be enough in a year's time to differentiate their product or company.[10] Since the machine would be challenging for a more established company to develop and produce in a year, it seems that the original Robo3D product design may not be competitive by the time it becomes available. According to the update, Robo3D is working through these concerns to determine the new company's greater contribution to 3D printing. There is no word on a production schedule for the Robo3D printer at this time.

Another crowdfunded 3D printer that found itself on a difficult journey to production was the Form 1 by Formlabs. A campaign for the Form 1 was launched on Kickstarter in September 2012 by three researchers from the MIT Media Lab. They developed an SLS 3D Printer for home use, which would be a first time this technology would be available in a small-scale printer. According to the group's Kickstarter campaign, they decided to develop the prototype because current FDM (FFF) printers are expensive and do not "meet the quality standards of the professional designer."

Form Labs asked Kickstarter for $100,000 to bring their design to production. The online community responded with a staggering $2,945,885 raised from just over 2000 individual backers. The majority of these pledges were made at the hefty $2500-$3000 pre-order price as 3D printing enthusiasts and tech blogs gushed over the machine's sleek design and high-quality results (the Form 1 promises a layer height as small as 25 microns).

Formlabs and Kickstarter were hit with a lawsuit in November 2012 by 3D Systems, which holds the patent for the SLS printing technology promised in the Form 1. According to a press release from 3D Systems, the lawsuit

10 "Update at RoBo 3D Headquarters!," Robo 3D - Kickstarter Update (blog), April 29, 2013, http://www.kickstarter.com/projects/1682938109/robo-3d-printer/posts/467427.

is "seeking injunctive relief and damages for infringement of one of its patents relating to the stereolithography process."[11] It is unclear whether the ongoing litigation has had an effect on Formlabs or its ability to deliver the Form 1 as promised.

The uncertain future of the Form 1 quickly became a talking point within the 3D printing community, which welcomed the idea of a desktop SLS printer to the market. However, with forum discussions asking if anyone had spotted a Form 1 printer working in the wild, it seems that some onlookers may have begun to doubt the odds of the machine becoming a reality in the face of the lawsuit. According to Formlabs website, there is no cause for concern. The company states that pre-orders for the Form 1 printer will ship in November 2013 and the company has not indicated any change from the machines original SLS design.

Pre-ordering the latest tech products is an exciting aspect of Kickstarter, but the crowdfunding process is about much more than a marketplace for the yet-to-be-developed. Some product ideas have changed the conversation around 3D printing with applications that capture the imagination of a broad audience. One such product is 3Doodle, a 3D printing pen. 3Doodle raised more than $2 million for its design, which puts the hot end and extruder of an FFF 3D printer in a handheld drawing tool. Artists can draw or trace designs in plastic and then lift the sketch off the page to create in three dimensions.

Two more crowdfunded machines that have expanded on the concept of a 3D printer are the RigidBot and the DeltaMaker. Both designs attracted attention for

11 "3D Systems Announces Filing of Patent Infringement Suit Against Formlabs and Kickstarter," 3D Systems - Press Release (blog), Novembre 20, 2012, http://www.3dsystems.com/fr/press-releases/3d-systems-announces-filing-patent-infringement-suit-against-formlabs-and-kickstarter

rethinking how a 3D printer is designed. The RigidBot is an easy-to-assemble printer kit that features an expandable frame that allows the machine to accommodate a variety of sizes and shapes. Similarly, the DeltaMaker is the first desktop printer concept to use a "delta" configuration (see Chapter 8) for the X, Y and Z axis. The result is an attractive printer that is as impressive to watch in action as are the objects it creates.

It is difficult to know whether these funded innovations, or the countless unfunded ideas, will ever become real consumer products. However, the marketplace created by Kickstarter and other crowdfunding sites continues to fuel rapid innovation in 3D printing and related products and services. These communities are perhaps the best place to observe the energy and excitement that surrounds 3D printing technology.

5. Modeling Digital Objects

So you have an idea for something to make. Maybe you had another one of those thoughts like, "Why doesn't someone just invent a better...?" or "Wouldn't it be lovely if this thing had a...?" With access to a 3D printer, you can now turn that idea into a real thing.

Everyday inventors can manufacture their own creations at home or in a community makerspace instead of a factory. This eliminates the need for a costly, difficult process just to produce the banana slicer you dreamed up over breakfast. And with a 3D printer, you don't need to make 10,000 banana slicers to afford the endeavor.

Just like printing on paper, having a working machine is only one part of the process. To print a document, you have to design it first. This is done through word-processing or graphic design software. Then, before printing, the document usually needs to be adjusted to print correctly on the page. The same type of process is involved in 3D printing, but instead of pages and text, we are working with digital shapes.

3D printing begins with building your idea in specialized programs called *computer-aided design (CAD)* software. CAD applications are able to build objects in three-dimensional space instead of the flat, 2D documents with which we are more familiar. You can think of CAD design as a drawing program for real things, and many of the most popular options provide a graphical interface similar to 2D painting and design software you may have already used.

When setting out to turn your idea into a real, 3D-printed object, there are a few things to consider. First, is your idea physically possible to build? Additive fabrication (a.k.a. 3D Printing) is an amazing technology that can create almost any shape. However, there are some things that just can't be done based on the physics involved. This is because 3D printing is done one layer at a time, and each layer needs to cool and harden in place. This process works great in most cases, but features like overhangs or bridges (where a portion of your object is suspended with no support) may not hold up.

Once you are sure your idea will be possible to design and print, you will begin the process of modeling the object. This step can be as much fun as printing, where your ideas take on a real shape. Even if you're new to 3D modeling, you will find a variety of options for getting started right away. As your skills improve, there are intermediate and advanced CAD tools available that are capable of building complex items to exact specifications. Each new tool will require some learning–a few will require a lot of learning–but nothing about 3D modeling is impossible. The only limit here is your time and imagination.

Design Software

As with other types of computer design work–like photo editing or graphic illustration–there are a variety of 3D modeling programs available for the task. Since 3D modeling and CAD software makers have only recently begun to focus on the consumer market, many of the leading applications can be a bit overwhelming to users without a background in engineering or design. However, a new crop of simplified tools is entering the scene, and some recent apps have been designed with 3D printing in mind. This makes basic CAD more accessible than ever to hobbyists and anyone interested in designing in three dimensions.

The *123D Suite* of software by Autodesk focuses on making 3D creation simple so anyone can begin designing computer-generated models for printing. The software suite consists of a variety of applications with capabilities ranging from 3D scanning to character design. This package is great for first-time users, and the features found in the 123D apps are excellent, yet the software is surprisingly offered as free downloads.

123D Catch puts the power of 3D scanning in the hands of anyone with a smartphone. With this app, anyone can create a 3D object by "scanning" an item in the real world. This is done by taking a series of photos from all sides of the object. The software translates these photos into a 3D model using reference points in the photos to gauge depth and separate foreground from background. A mesh cloud, or geometric representation of the model, is then made from the shape in the foreground.

123D Creature features a library of the torsos, eyes, extremities and other "monster parts" you can use to create fantastical animals from your imagination. These creatures are optimized for 3D printing. Along with *123D Sculpt,* which offers a middle ground between elementary school art class and more advanced modeling, users can sculpt digital models from a virtual block of material. When finished, the sculpture can be printed, making this a great tool for artists. Both 123D Creature and 123D Sculpt are also excellent for use in schools or at home with kids.

Rounding out the 123D suite is *123D Design,* which can be thought of as a friendlier version of traditional CAD software. This simple design studio introduces users to more advanced concepts like 2D extrusion and beveling, while designing objects with the push-pull interface of more complex programs like AutoCAD and Solidworks (more on these later in this chapter).

When you're ready to wade into a bit more serious modeling, *TinkerCAD* is a good starting point. This Web-based design app offers basic CAD features like shape building from cubes, cylinders and spheres. The drag-and-drop interface makes this tool approachable to anyone who has used a graphics editor or drawing program. When you have finished building your object, TinkerCAD offers an export option for 3D printing.

TinkerCAD had become quite popular among makers and 3D printer owners since its initial launch in early 2011. Users enjoyed its easy-to-use interface, integration with popular websites and team collaboration features (paid version). In March 2013, the company announced it would be closing TinkerCAD, causing panic among its users who would be left without a comparable alternative. Fortunately, AutoDesk purchased the tool a few weeks later, keeping the doors of TinkerCAD open for now.

For intermediate users there is *SketchUp*, a Google product, which was originally intended for designing architectural models, but it can be used to model a wide range of objects for 3D printing. There are a number of tutorials available from SketchUp team and community of users, which makes it a great option for someone looking for a tool with a lot of learning resources.

SketchUp's interface is much less intimidating than Blender or other comparable programs, however, it does not make for a great beginner's tool. It has a fairly high learning curve and takes more of a classical engineering approach to modeling than compared to the intuitive 123D suite, for example. The professional suite features a set of tools focused on architecture.

Also for intermediate users, Autodesk offers yet another solid tool called *Inventor Fusion*, which is a scaled-back but useful version of the more sophisticated modeling programs for which Autodesk is best known. Inventor Fusion is an immensely capable tool for modeling parts, components and other assembly essentials. It also includes solid and surface modeling abilities, along with translators and exporter packages for professional-level CAD tools.

You may need to do a little learning as you go with Inventor Fusion, so be prepared to do some Googling on your own. The tutorials, information and FAQ for this new software are sparse and a bit repetitive. Still, Inventor Fusion is a solid tool worth learning if you are a hobbyist or pro.

Popular with jewelry designers as well as architects, *Rhino* is a high performance tool for modeling vertical structures. The Rhino interface is similar to high-end engineering programs, with hints of SketchUp's architectural leanings. This software is worth a try for anyone looking for a capable modeling program, but it comes in at a hefty price tag compared to more capable software.

For quick modeling on the go, *Shapesmith* is a Web-based program for making, editing and storing 3D models directly in your favorite Web browser. Shapesmith is great for quickly building parts, prototypes and concepts for printing, making it a good tool for any 3D printer owner. Because it is a browser-based application you can access it from anywhere, but this also limits the scope of its technical capabilities.

If you are looking for a free CAD application with

advanced features, *Blender* is an open-software application famous as much for its large and active community as its intimidating interface. Blender is celebrated for having a very 'lean' interface focused on workflow and efficiency. An example of this is the number of single-key keyboard commands used for a fast and powerful workflow. However, this comes with a very steep learning curve. Blender is arguably the most powerful free package available, with python scripting capabilities and a large library of community-developed plugins. Blender can be tweaked to meet just about any modeling requirements.

For experienced designers, Autodesk makes some of the most popular CAD software in the world, and *AutoCAD* is their premier product. AutoCAD is a robust and powerful software suite with all the tools a modern engineer would need. These are the same applications that industrial and mechanical engineers are using professionally worldwide. AutoCAD has a number of support-based Internet communities, which is partially due to the software's very high barrier to entry. Learning to use AutoCAD requires a lot of patience and an inclination for math, so this tool is only for the most serious or experienced modelers.

Lastly, for our Windows-only friends, is the popular *SolidWorks* suite. This software is taught and used in classrooms and hackerspaces all over the world. The original RepRap printer parts were modeled using this software. It is much more affordable than AutoCad and has a large fan base with a number of communities offering support and knowledge.

Parametric Modeling

Another, very different method of 3D modeling is called *parametric modeling*. Unlike CAD software, parametric modeling is typically done using a descriptive language, or code, similar to computer programming languages used in software or website development. Parametric models are built using equations that form an object from a combination or basic shapes like cubes, cylinders and spheres. This method is used in industry for communicating specific attributes about the object during the manufacturing process, such as customization options.

Learning to design objects using parametric modeling is a very useful skill that we highly recommend for anyone serious about learning to work with 3D objects. The text-based approach may seem unfamiliar to those who have not worked with code before, but knowledge of parametric modeling will allow you to design objects much more precisely than the majority of graphical CAD tools. In addition, the ability to make customizable items is very powerful.

We recommend spending time to learn *OpenSCAD*, a free parametric modeling application. Instead of drawing a model in a graphical interface, OpenSCAD is based on code for plotting points and building shapes in 3D space. The goal is to build a solid object by combining shapes through a process called Constructive Solid Geometry (CSG).

OpenSCAD is widely embraced and supported by the open-source community, and has a deep set of resources for learning. Open SCAD is appropriate for beginners and experts, and it supports the most common 3D design formats (DFX, STL and OFF).

Preparing a Digital Model for 3D Printing

When you've finished designing your 3D object, you will then need to prepare it for printing. A 3D model on your computer, exists only as a set of digital information. To be printed, the model must be sliced into layers and then converted into instructions that the printer can read. These instructions are known as G-code. The entire process is appropriately called *slicing* and is accomplished by a software called a *slicing engine.*

The position of an object on the build platform can have a dramatic impact on the results of the print. It is important to remember that 3D printing is done in horizontal layers, building the object from the bottom up. In this configuration, fewer horizontal layers make for a stronger and more resilient object.

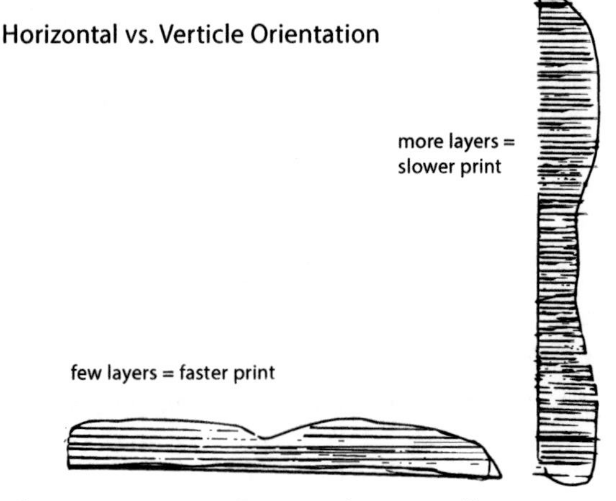

The orientation of some objects will not impact printing results. For example, a cube would be the same regardless of which side it was printed on. However, a plastic butter knife would turn out quite differently if printed vertically. In the vertical position, hundreds of layers would be needed to reach the tip of the knife. The resulting item would be more rigid and potentially more breakable than the same knife printed horizontally with far fewer layers.

In addition to position and orientation, the shape of some models can present unique challenges. Some objects may have difficulty adhering to the build platform because of their design features. For example, a rocketship sitting on its base will begin with tiny points where the fins touch the platform. This will likely cause the model to detach from the platform early in the build. This problem is solved by printing *raft* material, which is added around the tiny parts at the base of the model to aid in adhesion to the platform.

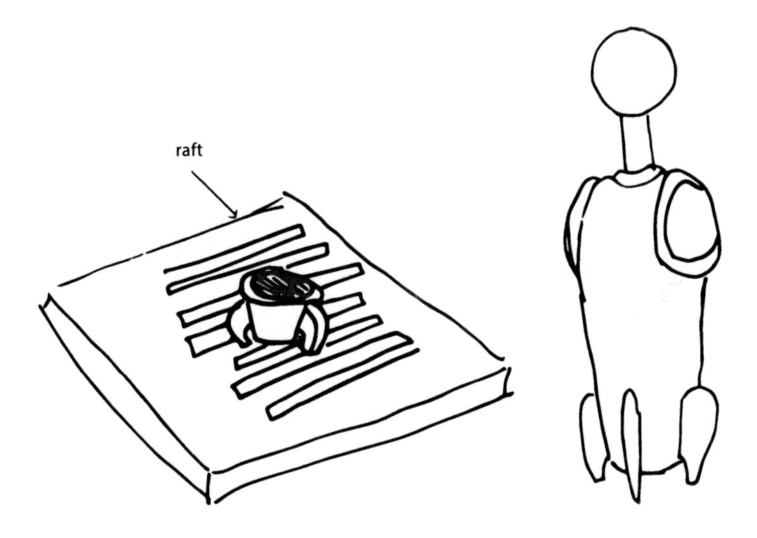

Models with dramatic overhangs may also have difficulty printing correctly. As a general rule, overhangs (unsupported structures) at more than 45° should not be printed without *support material.* Slicing engines include an option for adding support material, which is used to hold up overhangs in the model. Support material can be printed in PVA, which dissolves in water, or in a mesh of PLA or ABS that can be peeled away by hand.

To help visualize the need for support material, let's consider printing the following models–two different Pinocchio figurines. The first model is of Pinocchio as he normally looks, a well-mannered wooden boy. The second is Pinnochio after telling a lie. If printed in the upright configuration, the second Pinocchio model will fail because of the extended nature of his nose. At more than 45° of overhang, a 3D printer would not be able to build layers suspended in this way.

No Support Needed

Support Needed

The 45° rule may seem like a limitation to what is possible with 3D printing. However, printing even 45° overhangs gives additive manufacturing the ability to produce working objects that could not be manufactured through other methods. Gears and parts with moving, internal components can be built together with a single print job.

There are a number of other considerations and adjustments that can be made during slicing to help you get the best out of your 3D prints, regardless of your chosen machine or material. These settings can be a bit obtuse at first, so we will take a closer look at the meaning of each, as well as recommended settings in some cases.

Resolution

You will certainly want to consider the *resolution* of your print before producing your finished object. Much like a digital photograph, 3D-printed objects can be made in several different resolutions. We know that high-resolution photos use many more pixels, which are very densely packed together so that our eyes see a clear, smooth picture. Likewise, lower-resolution photos use fewer pixels, which causes the picture to look blocky. This is why photos from an older digital camera or cell phone look blurry compared to the higher-resolution of an iPhone or professional camera.

In 3D printing there are also high-resolution objects and low-resolution objects. A high-res object uses very thin layers (some can be thinner than a sheet of paper) to create a very smooth object. With high-resolution printing (100 microns layer height or less) it becomes difficult to see individual layers in the object. Low-res objects are made of fewer, thicker layers. These objects feel rough to the touch and contain layers that are more visible to the eye, like sediment or the rings of a tree.

Items intended for display purposes are typically printed in high resolution, while quick prototypes and everyday objects can usually be printed at lower resolutions and at faster speeds. It can be very challenging to get high-resolution prints from a consumer 3D printer. High-resolution objects take longer to produce than low resolution ones. This is because for every one layer of a low-resolution object, there could be 5 times as many layers in a high resolution object. Each of those layers is extra time spent printing, but they can make a substantial difference in visual quality. Also, a number of environmental factors, such as air temperature and filament material, as well as mechanical quirks, can prevent most current printers from turning out high-res prints on a consistent basis.

The number of layers in a print, or the object's resolution, is determined by the *layer height* setting, which is adjustable at the time of slicing. Layer height is measured in *microns* (one millionth of a meter). A model sliced with 100-micron layers would have a very high number of layers. The same model sliced into 340-micron layers would have a significantly lower number of layers. This is why a high-resolution print will take longer than a low-resolution version.

You may have already noticed that similarly shaped objects sometimes have very different properties. One example is a plastic drinking cup, which might be made of very heavy plastic and feel more like a ceramic. Other plastic cups, such as red party cups, feel light and flimsy. The range of weights, sizes and flexibility of plastic is what makes it a popular material for so many everyday things. 3D printing in plastic allows us to leverage these properties.

Not every plastic object can be made of the same thickness or weight. Otherwise, many of the unique plastic items we enjoy–from furniture to fly swatters–would not be possible. You can adjust many of your object's material properties through the custom variables and settings in the slicing software, including two important settings: shells and infill.

Shell & Infill

If we think about each layer of an object as a two-dimensional drawing which will be traced by the X and Y axes, then *shells* are the number of times the printer will draw each layer. The more shells on an object, the stronger it is. However, adding shells will also increase the print time significantly. Shells are also referred to as *perimeters* in some software and documentation. Use fewer shells when prototyping or printing decorative objects, use more shells when printing items that will be put under more stress.

Infill is the material used to fill the empty space inside the shell of an object. More infill will make an object stronger, heavier, and slower to build. Likewise, less infill is lighter and quicker to build. This is similar to traditional modeling and casting arts, where molds can be made solid (filled with material) or left as a mostly hollow shell. This is common in candy making, where large chocolate pieces are often hollow to cut back on cost and waste.

Infill is measured by percentage, so an object printed at 100% infill will be 100% solid. A 3D printer can extrude infill in several patterns. Some slicing engines create a grid pattern while others will use hexagonal or other geometric patterns. Items printed for display purposes rarely need more than 10%-20% infill, but functioning mechanical parts and pieces that will take more abuse will need 75%-100% infill.

A best practice is to add fewer shells and less infill on test objects. This will help save you time, plastic and money while refining your designs and settings for optimal printing.

Slicing Applications

There are several slicing applications available. Some are machine-specific, but other options work across a range of printers. Most commercial 3D printers have included software or recommendations for open-source applications.

Skeinforge is the most trusted slicing engine, but in its native form this open-source project is only for the most experienced computer users. Skeinforge is a tool chain made of python scripts that cut .STL files into horizontal layers, generate toolpaths to fill those layers with plastic and calculate the amount of material to extrude. Consumer interaction with this powerful tool is most likely to be through one of many popular slicing programs using Skeinforge as a slicing engine.

One such example is *Cura*, which borrows heavily from the Skeinforge source code but with a few key differences in programming logic. Many claim Cura to be faster than *Skeinforge*, although our tests show average speed when slicing. One of the best features of Cura is the ability to toggle from advanced to beginner settings. This is great for anyone just learning how to work with the different parameters. Cura is available for both Mac and Windows, and offers a deep level of G-code customization for advanced users.

ReplicatorG is a highly customizable slicer that is very popular with Makerbot owners, although compatible with a number of 3D printers. In addition to slicing STLs into G-code, it features a full *G-code editor* which can be used to manually adjust, edit and tweak the output to exact specifications. ReplicatorG also features a *G-code sender* which allows the software to run jobs and update a printer's firmware over a serial connection.

Keep It Simple Slicer or KISSlicer is an STL to G-code slicing application focused on additive printing. Compatible with a wide variety of machines, KISSlicer is popular for its straightforward tools and drawing clear tool paths for the print head. The KISSlicer Pro package features extended adjustments and settings to add extra infill every (n) layers, as well as support for multiple print head extrusion.

Slic3r is a newer product that is very proud of its "from scratch" roots. The tool is packaged with a host of other popular printing software as Slic3r tries to provide an end-to-end solution for slicing and printing. The software is free, but its developers accept donations to help fund the project.

Once a model is exported as an STL file it becomes difficult to edit or change. *Netfabb Studio* is an application designed specifically for repairing and adjusting STL files. This can be a handy tool for viewing, editing, or analyzing STL and or other file formats. While it is free and optimized specifically for 3D printing, it is not open source.

Another proprietary slicer is Makerbot's *Makerware*, which works only with Makerbot's line of 3D Printers. Makerware uses a slicing engine called MiracleGrue to prepare models for printing. Files exported from Makerware can only be read by Makerbot machines. This can be a problem for advanced users who will find the features and performance limiting. However, some Replicator owners replace the proprietary Makerbot firmware on their printers with the open-source *Sailfish* firmware, allowing the printer to work with the other options covered here.

Makerware does offer users some very nice features, such as the ability to drag and drop models on to the build platform. Still somewhat early in its development, Makerware has its bugs and quirks. The software particularly struggles with high-resolution slicing.

Some slicing applications like Makerware and ReplicatorG feature the ability to connect to the printer from your computer to run and monitor jobs, there are also independent applications that perform these functions. These programs are essential for printers without an SD-slot or onboard job management.

Two options here are *Pronterface* and *Repetier-Host*, which are used for sending G-code commands to the printer. The software interfaces with the printer over a serial connection, sending instructions as it works. Repetier also has options for web server management and queuing jobs for multiple machines.

6. An Internet of Things

When objects become digital, there are a number of advantages. We already know that an object can be modeled on a computer using CAD software. From there, that digital model can then be adjusted, scaled, duplicated or changed in infinite ways without consuming the time or resources it would take to do the same work in the physical world. We enjoy this advantage every day with documents we create at work and home. Imagine if we had to do typesetting on a printing press, or even write with pen and paper, every time we sent an email or posted on a friend's Facebook page.

Just as digital text and imaging have led to websites and social networks where we can easily share information, digital objects will lead to new and interesting possibilities. If an object can be digitized, it can be copied, sent, downloaded and shared. We are used to this with text, images and video, and these activities have become daily occurrences for many of us. We now get as much of our news, information and entertainment from the Internet as from television and print. The same may be true one day for physical objects.

Just like an Internet trend–such as a viral video–is created, remixed and spread by many people, so too are digital objects. What may eventually be a real thing (or many real things) could start out as a model designed by an inventor in Iowa, remixed into a derivative item by a graduate student in India and later sold by an entrepreneur in Australia.

This is the very same "Internet effect" that has revolutionized (or disrupted) so many once-physical industries–music, films, print media, retail stores–and brought us new products, even new markets, that were not possible before going digital. Physical products like the Pebble smart watch, television shows like Arrested Development, and musical performers like Justin Bieber and Carly Rae Jepsen owe a great deal of their initial success and continued support to this phenomenon.

The Internet effect has already begun around digital objects and 3D printing. This was never more clear than in the case of the Elevation Dock, an iPhone dock from Portland, Oregon's Elevation Labs that raised nearly $1.5 million on Kickstarter in February 2012. The attractive, aluminum dock was marketed as "the dock Apple should have made," offering a style, weight and elegance that surpassed the light plastic design sold by Apple for nearly half the price of the official version.

Although the product was a crowdfunding success, the company faced a major challenge when Apple released the iPhone 5 only weeks before the Elevation Dock was to hit the market. The new iPhone is equipped with a redesigned connector that is 80% smaller than the one used on previous Apple devices. This was the first time Apple had changed the connector on the iPhone since its initial launch in 2007, and the unexpected move would mean that new Elevation Dock owners would not be able to use the product with the iPhone 5.

Only a few years ago, this would have been a major disruption for a small business relying on slow, expensive prototyping and fabrication methods to retool. However, that is not the story of the Elevation Dock. In a matter of days, the 3D printing community responded with a user-designed adapter that would fit into the product, making it compatible with the iPhone 5. Elevation Dock owners with access to a 3D printer could download and print the part, upgrading their dock for free.

Elevation Labs quickly followed suit, offering a similar adapter for customers of the original dock while designing an updated model of the product. Without the ability to produce rapid prototypes, the 3D printing community would not have been able to assist in keeping the Elevation Dock afloat.

New fabrication techniques also make it possible for companies like Elevation Labs to respond quickly to the changing market. As 3D printing expands, we are likely to see the effects of digital objects in more situations like this, with some products eventually being sold and distributed in digital form.

Digital distribution of objects is an exciting idea in most cases, but there are some potential applications that may give us pause. Printable weapons, including handguns, have been a controversial topic in the mainstream, dominating the public discussion of 3D printing at this early stage. As the United States debates gun ownership rights, the ability to print working, plastic weapons is welcomed by gun advocates who see 3D printing as an expansion of the Constitutionally guaranteed right to firearms. Others fear that 3D printing will lead to untraceable weapons with the potential for criminal use.

After months of speculation, a 3D-printed gun was fired for the first time on video and shared on YouTube in early May 2013. The files needed to print the gun were also made available for download. Media attention quickly followed, and the American conversation about 3D printing centered around guns. In response, the United States Department of State requested that the digital files be removed from the Internet, although they remain available via Bittorrent file sharing.

The current state of 3D printing offers challenges on both sides of the gun issue. On one hand, opponents fear widespread access to unregistered (or worse, undetectable) guns. Supporters believe that if a gun *can* be made with a 3D printer, then the right to do so is guaranteed in the U.S. Constitution. However, consumer 3D printers are not quite capable of reliably printing a handgun using PLA or ABS plastic. These complex, moving parts (not to mention the explosive force involved in firing a weapon) require some very exact specifications.

While it is still possible to print a handgun using a 3D printer, the time, cost and failure rate involved does not yet make this a worthwhile endeavor. However, that will soon change as printers become more efficient and affordable. This means the time to discuss digital guns is now, while the technology is not yet mature.

Regardless of your intended use for the technology, 3D Printing can be tricky or costly in a number of cases, with printer manufacturers working to advance the capabilities of their machines while keeping them somewhat affordable to consumers. This means that owning a 3D printer is not yet in the cards for some, especially those whose work only requires occasional model making, prototyping or crafting. To address the casual user some companies are leveraging industrial quality printers to produce one-off orders for 3D printed objects.

Shapeways has created an online marketplace for 3D-printed goods based on this idea. With Shapeways services, a digital model can be printed, shared and even sold without ever owning a 3D printer. Instead of printing at home or in the office, users can upload digital files of their designs to Shapeways where they can be fabricated using a variety of materials.

To accomplish this, the company operates a "3D printing factory" in Long Island City, New York, where designs are received digitally, printed on industrial 3D printers, dyed or finished and shipped off to the customer. This type of service has its advantage of making expertly printed objects available on demand, or at least with two-day shipping.

Shapeways and similar services like Ponoko and Sculpteo have garnered significant interest and investment from those who see advantages in a marketplace based on digital objects. Beyond ordering custom prints, these sites offer the ability to set up shop for your designs. Much like Etsy and eBay, which have fueled niche markets around handmade and vintage goods, Shapeways serves as a destination for 3D art and creations.

The difference here is that there is no product or materials to stock or store, and no orders to fulfill for the seller. When someone purchases a Shapeways object, the item is automatically printed and delivered to the customer. In a sense, it is like making a factory available for everyone to produce and ship custom goods without the upfront cost or expensive equipment to buy.

This model is also being applied in a new service announced by office supply retailer, Staples.[12] The company will make custom 3D printing available for its customers, along with existing services like photo and document printing, beginning with a limited European launch in early 2013. Staples will use printers by Mcor Technologies, which form objects from layers of paper (instead of plastic filament) that can feature photo-realistic coloring. It is not yet unknown when this service will be available in the United States from Staples or other

12 Senese, Mike. "Staples announces in-store 3-D printing service." CNN, December 01, 2012. http://www.cnn.com/2012/11/30/tech/innovation/staples-3-d-printing.

retailers.

For a more Do-It-Yourself approach to fabrication services, *makerspaces* have begun to appear in cities and towns around the world. These community workshops, which predate consumer 3D printing, offer access to equipment like CNC mills, laser cutters and FDM printers, as well as tools and training to do your own welding, soldering, bending, sewing or drilling. Makerspaces typically allow open access for members.

TechShop began the first chain of makerspaces in the United States, with locations in the Bay Area and in tech centers like Austin, Texas. Focused on makers with ideas to build, TechShop quickly trains its members on the tools, technologies and processes they will need to prototype an invention. This is done through a series of workshops offered at each location.

TechShop and other makerspaces are possible because they pool the cost of equipment that would otherwise not be available to an everyday inventor. This was the case for Jim McElvey, who founded the mobile payment service Square with partner Jack Dorsey (Jack is also a co-founder of the social network, Twitter). McElvey prototyped the first Square credit card reader using milling machines and other devices available at the TechShop location in San Francisco.

TechShop is now expanding with new locations across the United States, including some in manufacturing cities like Pittsburgh and Detroit. Ford Motor Company collaborated with TechShop around its Detroit location with a partnership that would encourage use of the space for innovation. Ford employees who work on an idea that becomes a company patent receive a three-month membership to TechShop Detroit for creative pursuits.

If you are looking for access to a 3D printer or other fabrication equipment in your own community, you might also try your local library. Makerspaces are beginning to appear in some public libraries that see access to these tools as an important extension of a library's mission to provide information and education to the community. Library fabrication labs are not yet widespread, but they may prove to be an important step for communities in need of a makerspace, and for libraries looking to modernize the services they provide.

The Fayetteville Free Library, located in Fayetteville, New York, has been leading the way in this area, offering the use of 3D printers to its patrons. Fayetteville is a quiet suburb of Syracuse, an industrial-era city with a rich history of manufacturing. The area was once home to major manufacturers of minerals, metals and machines that were sold worldwide. Now the library is bringing small-scale manufacturing back to the region as the first library in the United States to launch a makerspace.

Fayetteville's "Fab Lab" is located in a factory once used by the Stickley Furniture company, which has been producing handcrafted furniture in the Syracuse area for over 100 years. Soon a different kind of manufacturing will be taking place in this historic space as desktop equipment ranging from sewing machines to 3D printers fill the floor. FFL plans to feature its original Makerbot Thing-O-Matic printer (Fayetteville is also believed to have been the first library to offer a 3D printer for patron use) as well as a Replicator and a new machine by 3D Systems. The library plans a variety of classes and other programming around this new resource.

Community and collaboration are also growing around 3D Printing through online sharing of digital objects. *Thingiverse* is a site developed by Makerbot Industries that is a cross between a repository of 3D models and a social network for 3D printing hobbyists. The site is free to use (even if you don't own a Makerbot) and contains thousands of community-created models that can be downloaded and printed, or turned into derivative creations.

Thingiverse has become one of the most active communities for 3D printing. This is largely due to its use of social networking features to promote sharing and engagement around digital objects. Each Thingiverse user has a profile where his or her models are displayed. If the user has printed the object in real life, a photo can be added along with descriptive information and instructions. Users can follow others, "like" their favorite things, and download unlimited models to print.

7. Selecting a 3D Printer

In Chapter 3, we covered the most popular 3D printer options, as well as a few promising examples that are making serious contributions to the industry. Many of these machines are worth considering if you are in the market for a 3D printer. However, you are certain to find a dizzying array of choices as you shop–from prefabricated printers to the countless kit machines available online.

As you can see, the young age of the 3D printing consumer market has not slowed the rate at which new machines are introduced. Combined with the lack of information available compared to other, more mainstream technologies like personal computers, the wide range of choices can seem quite complicated.

To help you decide which 3D printer is right for you, we have broken down the features and specs common across all options. By understanding these features–as you would the RAM and hard drive space of a personal computer–you will be able to make an informed purchase decision regardless of manufacturer or model.

Before looking at specific features, it is important to carefully consider the types of things you would like to make with your 3D printer. This is because, despite how blogs and magazines tout the futuristic capabilities of 3D printing, there is not yet a machine that can make anything on demand. Until we can work directly with molecules, we must settle for a variety of machines built to work with specific materials.

Will your machine be a tool to support a hobby or career? Will you use it just for fun? Maybe a little of both? The answer to this question will help determine everything from the style of printer and the material it uses to the speed and performance needed from the machine. Of course, a 3D printer can certainly be used for a number of work and fun activities, and a hobbyist will probably want to select a versatile option that will allow for some flexibility in use.

If you are looking for a workshop printer, begin by considering the physical design of the machine. What are the machine's dimensions? Some printers are quite large and will dominate a workspace or bench. Can comparable performance be achieved with a smaller footprint? Similarly, some chassis designs can get in the way of productivity. For example, consumer-oriented printers may not offer the ability to mount multiple filaments, change the filament during the print or easily access and remove the object. Cleaning and upkeep can become a challenge for a work machine.

On the other hand, artists, designers and schools may prefer a less industrial machine. Consumer options offer more approachable interfaces, or better support for the less mechanically inclined. If you plan to use your 3D printer for creative or educational purposes, you may not want to become a machine operator in the process. This will require balancing the performance you need against the amount of fine tuning the machine demands.

Regardless of your specific needs, there are common features to all Fused Filament Fabrication (FFF) 3D printers. We will consider each attribute in terms of their overall impact on the machine and its abilities, as well as what they mean to a professional, hobbyist and casual user.

Chassis design

Just like the people operating them, 3D printers come in many shapes and sizes. The term "desktop printer" can be deceiving unless you remember that the machine may indeed take up your entire desktop, and some do!

Some printers have an open body design with large windows in the walls. Others may not have walls at all, sticking with the bare frame construction. This type of chassis allows air to pass through, which can cause some challenges depending on the filament material being used.

Open-air designs are not ideal for printing ABS plastic because of its sensitivity during the cooling process. Machines that print ABS must have a build environment where the object can cool evenly over time. In the open air, the bottom layers of ABS will cool faster than the top, causing the object to warp. The effect of drafts and quick cooling are minimal with PLA (in fact, PLA should be cooled more quickly), which is why many PLA-only printers have an open design.

If you are planning to print professional parts with your 3D printer, then you may want the ability to print the more flexible ABS along with the rigid PLA. To get high-quality ABS printing, you will need to select a machine that offers a controlled environment in the build chamber.

This is sometimes accomplished with additional features such as closed chambers with doors and covers. These panels must be opened or removed to access the build platform. This will keep the air temperature more stable inside the printer, but a closed machine can become tedious in a production environment. Each completed print job will require the user to open the machine and remove the object. This does not seem like much at first, but repeating this task many times an hour can quickly add up. Cleaning the inside of a closed machine can also be difficult.

Hobbyists looking to explore a variety of applications for the 3D printer may choose one with a somewhat open chassis design, especially if the machine will be producing prototypes or artwork that does not need to perform to exact tolerances the way a machine part might. An open design uses less material, which often reduces the retail price of the machine or kit. Those planning to use the 3D printer at home, in school or for fun need not worry much about these details. A printer designed only for PLA will be a fine choice.

Another concern when deciding on a chassis design is vibration and movement while printing. The moving parts of the printer cause a lot of directional force and momentum, especially when printing at high speeds. To prevent the machine from wobbling about, a stiff, rugged chassis design is best. Chassis made of flexible material, such as acrylic (plastic) or wood, will jiggle much more. Loose joints and weak construction of more sturdy materials can also cause this type of movement, which will eventually shift the axes out of alignment. Even minor fluctuations can reduce the quality of a printed object.

Build volume

The *build volume* of a 3D printer is the total area inside the machine for printing objects. In other words, the build volume tells us how large the pieces we print can be. Some desktop 3D printers offer large build volumes capable of making nearly foot-long objects with a depth of 7 or 8 inches. These machines can make substantial objects as large as a vase or an adult shoe. Other printers have volumes half that size, capable of producing knickknack-sized things.

It is important to note that printers with larger build volumes can make one or many small items in this larger area, while smaller machines are not expandable. If you would like to make bicycle or car components, for example, a 3D printer with a larger build volume would be ideal. The bigger machine can produce larger parts as a whole, while still printing small items in batches of one or many. A smaller build volume would mean that some larger items would need to be printed in pieces and assembled, or that some parts could not be made with the printer.

On the other hand, someone who intends to make 3D-printed jewelry could be very productive with a smaller build volume. Because most of the pieces used for jewelry making are tiny or assembled, a smaller printer would be able to produce everything needed. If you do not plan to print large, whole objects, a printer with a smaller build volume of 5x5x5 inches will save you quite a bit on the price tag.

Filament Material (ABS vs. PLA)

When selecting between desktop 3D printer models, you may choose between machines capable of printing ABS plastic, PLA plastic, or both. Consumer-oriented printers often specify which type of filament material should be used with the machine. Printers capable of

ABS printing are typically able to print with PLA as well. However, PLA printers lack a heated build environment and sometimes the temperature range to properly print using ABS. Additional materials, such as PVA or Nylon, may be printable as well. You will want to check the recommended settings for each material against the specifications of your printer.

Professional users, or those planning to make flexible or complex parts, may choose a printer capable of ABS printing. ABS is a durable, flexible plastic with a higher melting point than PLA. It is therefore good for making containers to hold hot liquids, for example, or flexible items like joints. However, ABS has more sensitivities than PLA and requires additional components, like a heated build platform and more patience to get the print just right. Less experienced users may want to avoid the complexities of ABS printing.

PLA, which is starch-based plastic, is rigid and strong. It can be printed without a heated build platform and performs better when making tight angles. Where ABS will deform and warp when bent, PLA will snap. This makes it less appealing for making moving parts, toys and other items that need to stand up to outside forces. PLA has fewer sensitivities and can print in higher detail than ABS, making it a versatile material overall, and an approachable starting point for new users.

Single or Dual Extruder

The number of extruders (or print heads) on your printer will determine the number of materials you can use to make a single object. Some machines offer dual extrusion capabilities, meaning the machine can simultaneously use two separate materials. By combining different materials or colors, more complex objects can be created.

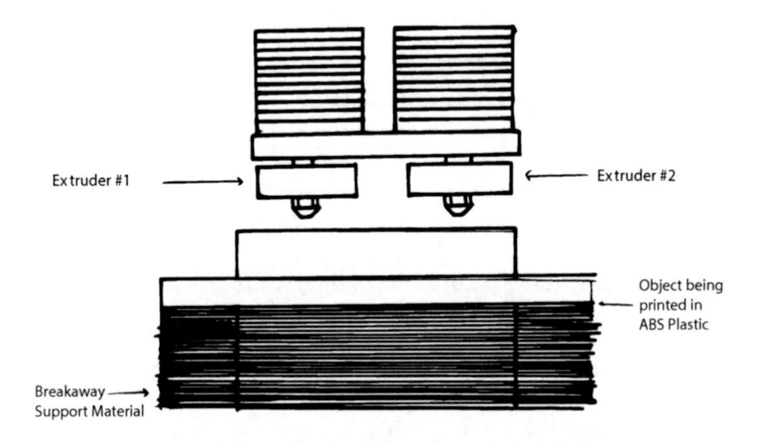

Another potential use of a second (or even third) extruder is to print dissolvable support material called PVA. In the same way two colors of filament can be printed together, PVA can be added to gaps and spaces to increase structural support during printing. Once the object is completed, the PVA can be dissolved in water. This can produce a more intricate object than possible with ABS or PLA alone.

Resolution and Speed

The resolution of a 3D printer refers to the thinnest layer height the printer is capable of producing. The thinner the layers of an object, the higher its resolution. In other words, to call a printer "high-res" is to say it can print objects using very thin layers, and therefore smooth surfaces where individual layers are difficult to see with the human eye. Likewise, a low-res printer (or setting) will use fewer, more visible layers that form a surface that is rougher to the touch.

REFERENCE CHART | Object Resolution & Layer Height Range

Object Resolution	Layer Height (microns)	Layer Height (mm)
High	25 - 150	0.025mm-0.15mm
Medium	160-250	0.16mm-0.25mm
Low	260-340	0.26mm-0.34mm

*A sheet of paper is about 100 microns, or 0.1 millimeter thick.

Because high-resolution objects have so many layers, they will take longer to print. How long exactly depends on the speed and capability of the printer, which is an important factor in making this choice. Printers that can work at speeds as high as 120mm/s will print much faster than those at 80mm/s, this can shorten the time it takes for a high resolution print by as much as several hours.

Some printers are also limited in their technical capabilities for object resolution. It's important to consider the resolution and purpose of the objects you plan to print on your machine. If you want to print high-resolution objects quickly, like a small business or workshop might, you will need a printer that can produce high resolution objects at faster speeds (120mm/s). If your high resolution objects do not need to print so quickly, then a more affordable but slower machine with high resolution capabilities is a much economical option.

The resolution of a 3D printer may not be the most important factor to some users. For example, a product designer who needs rapid fabrication abilities may find a machine optimized for reliable medium-resolution printing to be the best option. It is important to remember that high-resolution settings on consumer 3D printers (as small as 100 or even 25 microns) add complexity, and increase the potential that a print will fail. The default setting on most 3D printers–even the expensive ones–is usually medium-resolution.

Connectivity and Software

Most 3D printers use a USB connection or SD card reader to transfer jobs from software to the machine. A printer that connects only by USB will need to be controlled by a nearby computer or laptop. This is usually done through software that also manages the settings of the printer. Adding SD card capabilities to the machine allows for printing without need of a computer connection. Users can save their design files on an SD card (many laptops now include built-in SD slots) and transfer the file to the printer. There is usually an LCD display on the 3D printer that will allow for navigation of the SD card file system and control of the printer's functions.

If multiple users will be printing objects, or if you do not have a dedicated computer in your workspace, then a printer with an SD card option is well worth it. 3D printers with Wifi connectivity, such as the Cube printer by 3D Systems, are slowly entering the market as well. Wireless printers will offer another way to share and print to a workspace printer.

Community and Support

Since the desktop 3D printing industry is only a few years old, support for these devices is not always as plentiful as with more established technologies. In fact, it is possible that a new 3D printer owner could be one of only a few in an entire city or state. This means that the majority of information, support and learning around 3D printing is found online in user communities, or directly from printer manufacturers.

The best resources for 3D printer owners are still

found in forums of researchers, developers and hobbyists who gather on sites like Google Groups to share tips and ideas, or to answer questions from other users. These groups are particularly active around printers derived from the open-source RepRap project because community members are able to access detailed information about the machine designs to address issues or suggest improvements.

Direct support for your 3D printer is available from some manufacturers who have dedicated resources to phone and online channels for customers to address technical or operating issues. However, most 3D printers are made by small to medium-sized companies focused on developing and packaging the technology for consumer use. Some offer support documents and introductory material on their websites as well, but the content is limited and often incomplete. No matter how you plan to use your printer, be sure to select one that comes with the support resources you feel comfortable using.

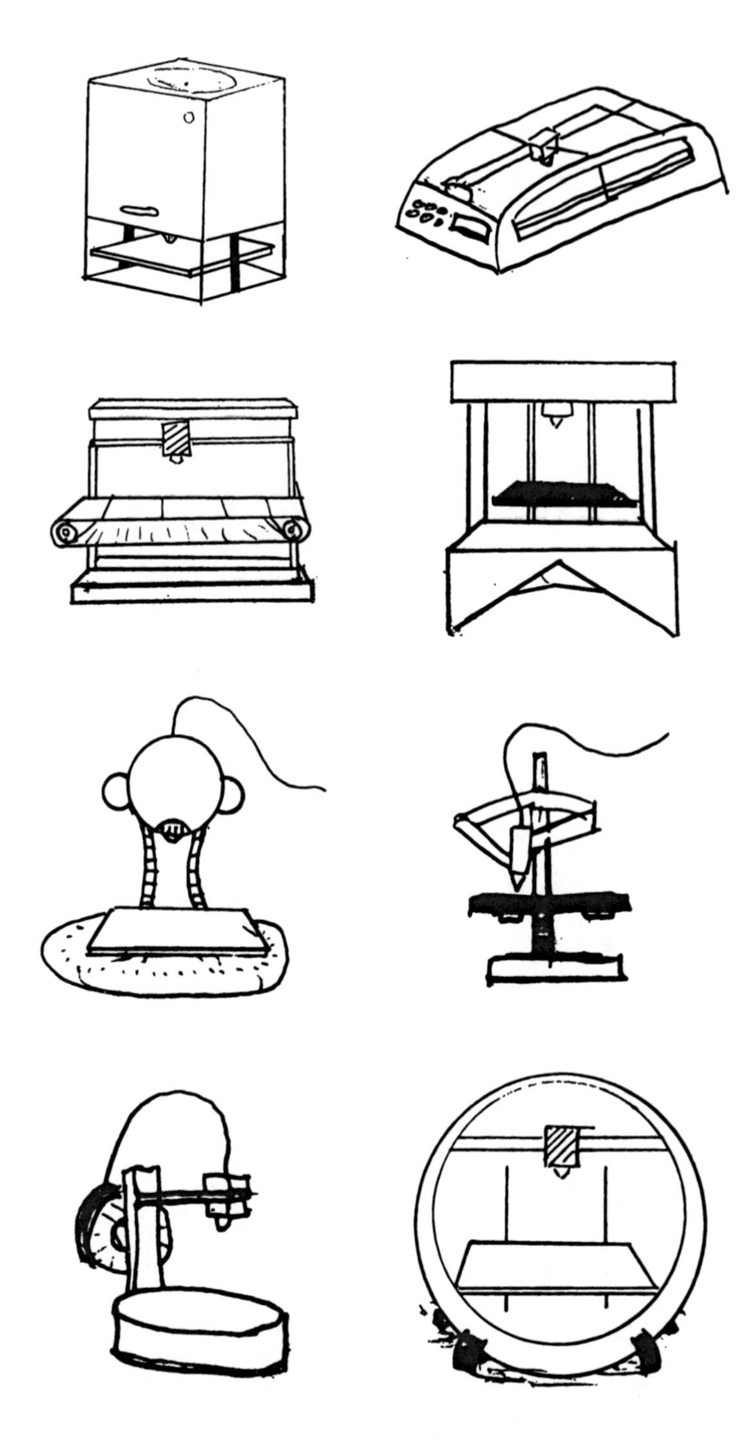

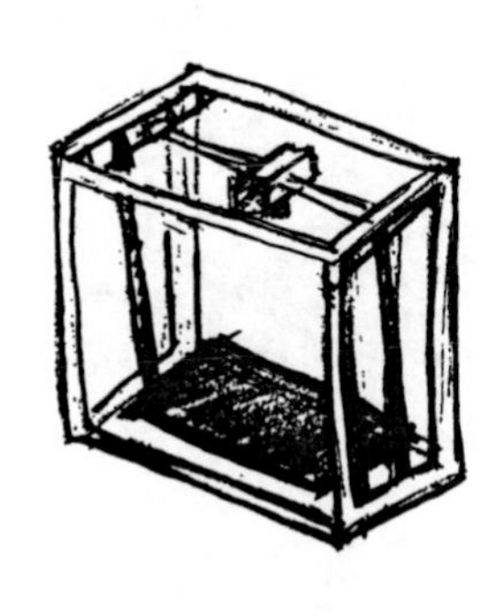

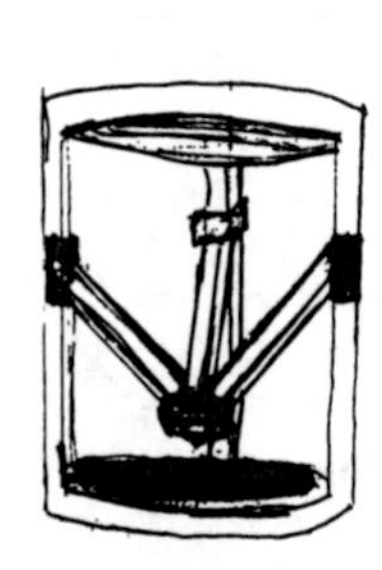

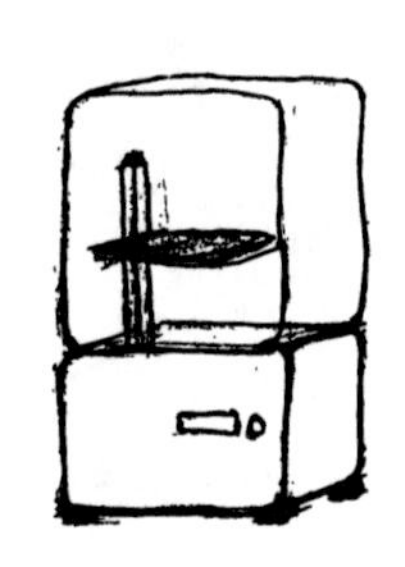
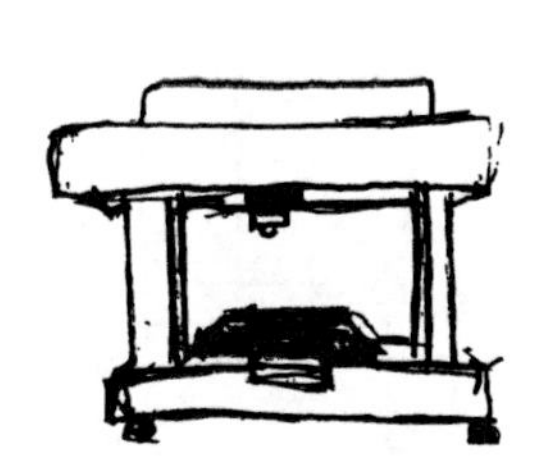

Part II: Inside a 3D Printer

8. The X, Y and Z Axes

At some point in school we were all handed grid paper and introduced to the concept of plotting a point based on (X,Y) coordinates. We learned that to plot the coordinate (22,33) on a graph we would take our finger, starting at the origin (0,0), and move 22 units across the x-axis. Then we move 33 units up the y-axis. There we plot our point.

Modern-day inkjet and laser printers work exactly like this. The files we print from our computers, like PDFs and MS-Word documents, get broken down into a bunch of (X,Y) coordinates that tell the printer where to move the print head, when to pull the paper further out and when to spray ink. Just like when we plotted the point with our finger on the X and Y axes, the printer is doing the same. Here, the print head moving back and forth is the x axis, and the paper being pulled through is the Y axis.

When we talk about something as "two dimensional," this is one way of saying that it can be plotted using two axes, which we call X and Y. When something is "three dimensional," it means it exists on three axes, or dimensions. As we already know, the first two axes are X and Y. The third dimension is called the Z-axis.

Imagine once again that you are looking at the same piece of paper from our first example, with the two-dimensional point plotted at (22,33) on the X and Y axes. If we were to plot an additional point on the Z-axis (i.e. add a third dimension), this value would tell us the distance that the point rises up off the paper. In other words, the Z-axis measures depth, while the X and Y axes measure the width and height of an object. Adding the Z-axis is what makes an object tangible.

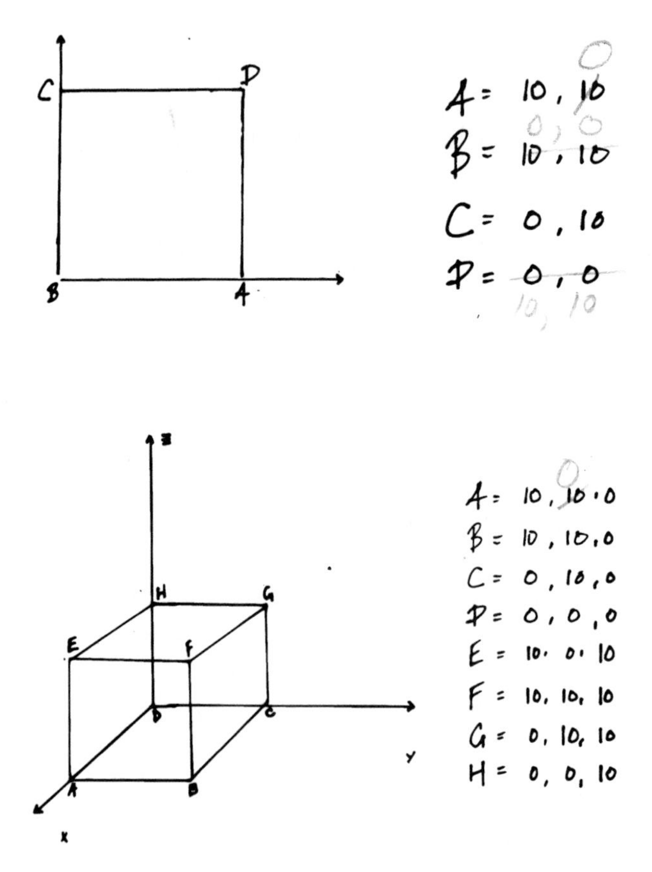

With an understanding of all three axes, we can now plot a three-dimensional point using the coordinates (22,33,44). These values refer to the position of the point on the X, Y and Z axes respectively. To plot the point, we simply start at the origin (0,0,0), move across the X-axis to 22, up the Y-axis to 33, and finally up off the page to a distance of 44 units.

This system of reference for plotting and finding specific points in three dimensions is called the Cartesian Coordinate System. 3D printers use the Cartesian System to build objects in three dimensions, plotting points in plastic as the object forms one layer at a time. Starting at the build platform, the print head moves along the X and Y axes as it draws the first cross section of the object. This is the first layer that will form the base of the object. After each layer, the machine moves to the next point on the Z-axis to begin drawing additional cross sections.

Belts, pulleys and rails... oh my.

In order to print on three axes, we need three planes of linear movement. In other words, our object needs to be built in three separate directions: flat across the build platform (the first two directions) and up off of the plate (the third direction). The movement across these axes is accomplished through several *linear motion* techniques that have become de facto industry standards. The three most common approaches to linear motion are a belt and pulley system, a ball-screw, or a rail and wheel configuration.

A *belt system* works by using a closed loop rubber belt with tiny treads, or teeth, running along the inside. This works much like a conveyor belt or the treads of a tractor. One end of the belt loops around a free-spinning pulley gear, and the other around a gear attached to a *stepper motor*. As the motor turns, the teeth catch on the pulley gears moving the belt in the desired direction.

In this configuration, moving components are attached to a carriage that slides along a stationary rail positioned parallel to the direction of the belt. Linear bearings are used to attach the belt to the carriage and move along the rod as the belt spins. This type of setup can be run very quickly, but the nature of belts and the number of moving parts make it the least precise method.

Linear motion can also be achieved with a *ball screw* setup. This works by attaching a threaded rod and nut to a motor. The motor turns, causing the nut to ride back and forth along the rod. In this setup, the carriage is attached to the nut, which moves back and forth along the rod as well. Although slower and more expensive than a belt system, screw-driven systems are significantly more precise. Most z-axis rigs use a screw-driven system.

A *rail and wheel* configuration is also a popular option. This approach works much like the system traditionally used to move railroad cars along a track. The carriage, powered by a belt system, rides smoothly on wheels across a series of rails. This is a common setup for linear movement on industrial machines across a variety of equipment, including 3D printers.

XYZ: One premise endless configurations

There is only one recipe for making coffee: Run water through coffee beans. It is a simple formula, yet there are so many different types of coffee makers. There are percolators, presses, and one-cup devices. Some slowly drip hot water over ground beans. Some machines work by heating water over a long time, and others work by recycling water back through multiple times. In the end, these are each methods of concocting the same recipe. The method one chooses is often a matter of taste, or a decision made after considering the benefits and downsides of each option. The same is true about the arrangement of the axes inside a 3D printer.

So far, we have taken a look at how a printer finds points in space, and then how it moves the mechanisms to reach those points. However, the movement of the X, Y, and Z axes can be configured in a number of ways using a seemingly endless combination of mechanisms and alignments. The end goal, much like the simple coffee recipe, is a 3-axis machine capable of producing a three-dimensional object. How this is achieved is a matter of preference, creativity and a balance of pros and cons.

The most common design found in 3D printers involves suspending the X and Y axes above a moving platform. In this scenario, the build platform only moves in one direction, up and down along the Z-axis. The print head moves along the X and Y axes, drawing each layer in place as the Z-Platform steadily lowers towards the bottom of the machine. This design is used by the Makerbot Replicator, Ultimaker and Type-A Machines, among many others.

This configuration has become a de facto best practice in the industry for a number of reasons. Among them is the decreased movement of the object being printed, since the build platform only moves up and down on the Z axis, as if on a car lift. In addition, the machine itself will experience less vibration with fewer parts moving back and forth. This will cut back on defects in printed objects.

Machines like the Makerbot Thing-O-Matic and 3D Systems Cube take a different approach by moving one of the horizontal assemblies (X or Y) to the Z-platform. In this configuration, the build platform moves both horizontally (side-to-side in one direction) and vertically along the Z-axis.

With this type of bidirectional build platform, the print head is attached to the top of the machine along the X axis, and zips back and forth doing its job.

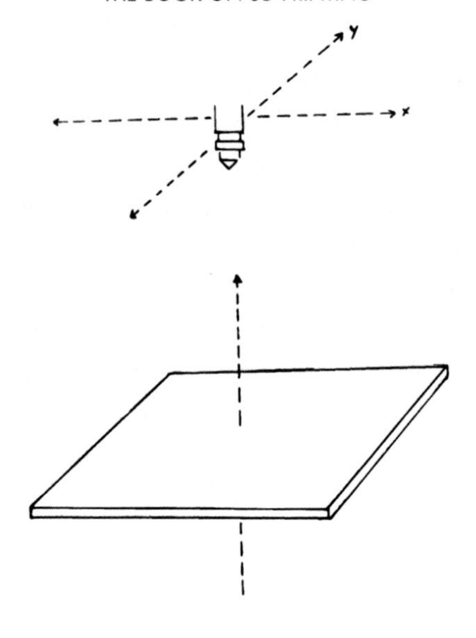

Unidirectional

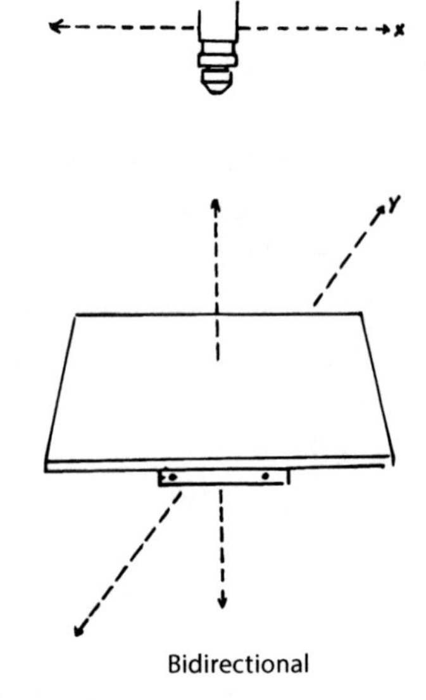

Bidirectional

Simultaneously the Z-axis assembly has an attached Y-axis assembly that allows the build platform to move back and forth horizontally as well as vertically. Because extra weight is placed on the Z axis to make the build platform movable on two planes, this setup requires a very sturdy chassis to perform accurately.

To picture a bidirectional build platform in action, imagine you are making an ice cream sundae from a soft-serve ice cream machine. You have a dish which you hold up to the nozzle where the ice cream is "extruded" from the machine. As the ice cream comes out of the machine's stationary "print head," your hands act like the build platform, slowly moving the dish from side to side to get a perfect, soft-serve swirl. By steadily lowering the dish, the ice creams builds layer upon layer until the sundae rises high above the top.

Less common than a moving build platform is a stationary platform design. In this scenario, the build platform stays in place while the print head moves on all three axes. Delta-style printers work in this way. A delta printer has three vertical rails, each with a carriage attached to an arm. The three arms meet at the center of the machine where the print head is hung. From there the machine moves each carriage individually, up and down, to articulate the print head around the build area. By moving one carriage up the rail above the other two, the print head will move in the direction of the carriage moving upward. Moving that same carriage below the other two, the print head will move away. Using this system the object is built up in place.

Instead of soft-serve ice cream, you can think of a stationary platform printer as working much like we do when decorating a cake with a tube of icing. Our arms would act as both the x and y axes as we moved them across the cake drawing out the birthday boy or girl's name. When we finish writing "Happy," and lift our hands up to move over and begin writing "Birthday," we have our Z-axis motion.

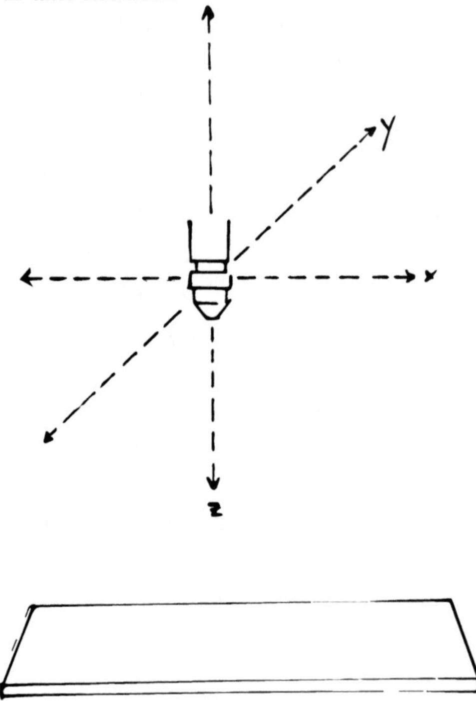

Stationary

Delta Mechanics

One way to explain delta mechanics is to imagine a swing attached to two posts. Unlike the swings at a playground, the chains holding up this swing are attached to the support poles with a hook we can position anywhere we like along the pole. If we put both of these attachments at the same point on each pole, the swing will be balanced in the center. However, if we raise one side and leave the other in place, the swing will move towards the raised side. Likewise, if we lower one side below the other, then the swing will move in the other direction. In this scenario we could move the swing from side to side as we find any position between the poles.

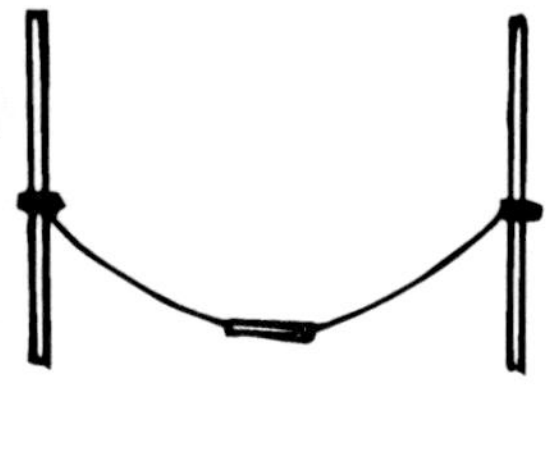

A delta printer works using this same mechanical structure. By adding a third support beam and attachment rope, or arm, we can articulate the print head to find any point between the three posts.

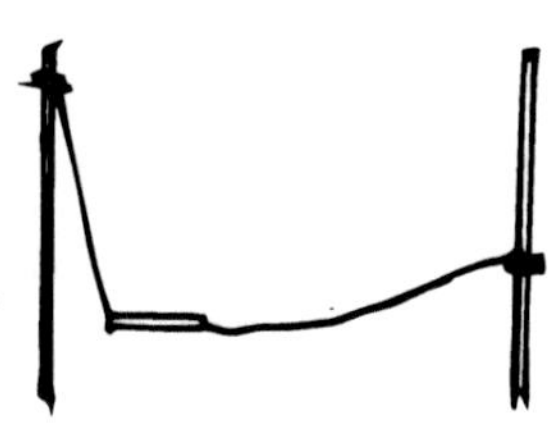

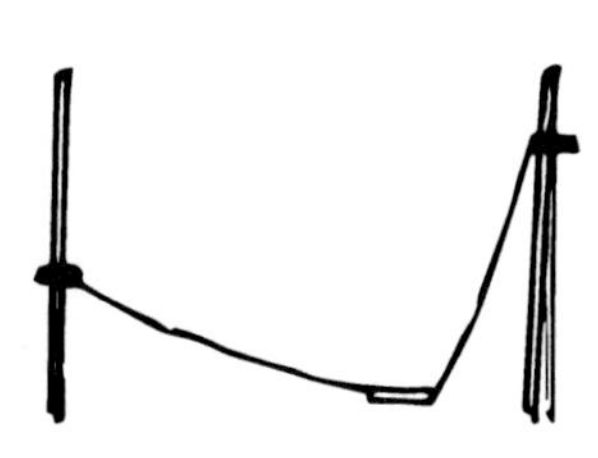

9. The Build Platform

The build platform, or bed, is the surface on which a 3D printer makes an object. Beds range in size from 2 in. x 2 in. to extreme machines with platforms several feet in size. Most standard desktop sizes range between a few inches and a foot in size.

The filament material being used in the printer will typically dictate the type of material used to make the build plate. Filaments like PLA and PVA will easily adhere to most surfaces at room temperature, whereas materials like ABS require a build platform that has been heated to around 80 to 100 degrees Celsius for proper adhesion.

Below is a quick reference showing which filaments may be used with three of the most popular build plate materials: Acrylic, Glass and Aluminum.

	ABS	PLA	PVA
ALUMINUM	⊗	X	X
GLASS	⊗	X	X
ACRYLIC		X	X

⊗ = HEATED

Heated solutions, specifically heated build platforms or heated build chambers, are used to combat curling of the object. Some filament materials (such as ABS) shrink quite a bit while cooling. However, some curling can occur with any material. As a 3D printer builds an object, the layers of the object cool at different rates. A heated platform or chamber works by keeping the lower levels of the print warm as the upper levels are added. This balances out the temperature gap and allows the object to cool evenly.

Of the common build plate materials, *Acrylic* is the most economical option. Unfortunately, the low melting point of acrylic plastic makes these platforms prone to warping. For this reason, they cannot meet the heating requirements for ABS printing. PLA will stick to an acrylic surface untreated.

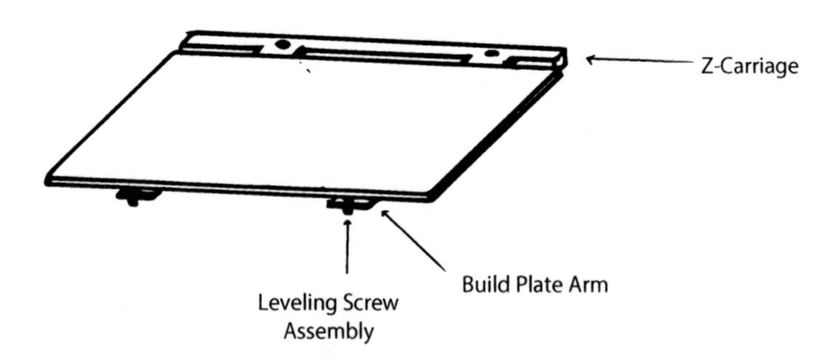

Glass offers a great performance-to-price value. Glass build platforms are extremely flat, but prone to cracking because of the temperature fluctuation of the machine. Borosilicate glass, made of boron oxide and silica, has a high tolerance for heat fluctuation and is the best option for printing ABS or PLA on glass. ABS and PLA will not stick to a glass surface untreated. (continue reading for print surface treatments)

Aluminum is the most expensive option for a build platform, but it offers some unique properties. Aluminum beds are most often used in heated platforms because they are conductive. Its precise machining capabilities also allow for the production of a very flat plate. ABS and PLA will not stick to an aluminum surface untreated.

Build Surface Treatments

Some filament materials need a little extra surface preparation in order to help the filament adhere. Some printer manufacturers recommend specific treatments, and a few suppliers even sell spray-on solutions. However, the 3D printing community has discovered a number of DIY treatments that help keep an object in place while it prints.

Painter's tape is one of the most popular solutions to combat curling because it works reliably in a variety of situations. As an object prints, the first layers of liquid plastic latch onto the fibers and wax in the tape, giving a strong adhesion to the build platform. Ideally, when applying tape to the build platform, you want to have the fewest and smallest seams possible between the rows of tape. Gaps and inconsistencies in the tape will affect the quality of the bottom of your object. Painter's tape works with most filament types, including PLA and ABS.

Hairspray is another popular solution. It is slightly less economical, but strong-hold hair spray has proven to be one of the best options for fighting curl. Lightly cover your entire build plate in a dusting of hair spray (AquaNet or similar bargain brand works well), allow it to dry for a few minutes, then print. This method also works with most filament types.

For PLA printing, a little *extra virgin olive oil* (EVOO) applied to a build surface is said to promote good adhesion and easy removal of finished objects. The starch-based makeup of PLA is what allows this unique combination to work. There are no significant benefits to printing ABS or other filaments on surfaces treated with

EVOO.

For a truly do-it-yourself option to prevent curling, you can try making some *ABS glue* to treat your build surface. ABS glue is a home-brewed concoction consisting of ABS pieces dissolved in acetone. ABS dissolves quickly in acetone to create a brushable solution that can be brushed on as an adhesive. When the liquid mixture is spread on a surface, the acetone will evaporate leaving a strongly bonded, solidified ABS in its place.

Some hobbyists will keep a jar of acetone on their workbenches to make use of scrap ABS pieces as glue. Extra acetone can be added to bring the solution to a proper consistency for spreading a light coating across the build platform. It is important to wait for the ABS glue to dry before printing. This treatment works with most filament types.

There are also several alternative build surface materials that have proven useful as new printable materials enter the market. 3D printing hobbyists are on a perpetual quest to find the ideal platform material that will work with the ever-increasing variety of filaments and printing methods.

Nylon, for example, requires a surface made from a material that is dense with cellulose for proper printing. Build platforms made from poplar wood or manufactured mediums like Garolite make a good solution. When considering alternative materials for the build platform, look for a very flat material with high temperature resistance to prevent warping and other issues related to temperature fluctuation.

Leveling

In order to print an object properly, the printer needs a flat and level surface to build on. Leveling a build platform is a bit more complex than just laying a liquid level across the bed and adjusting accordingly. This is because the build platform of a 3D printer must be leveled relative to the tip of the print head nozzle, not to the ground.

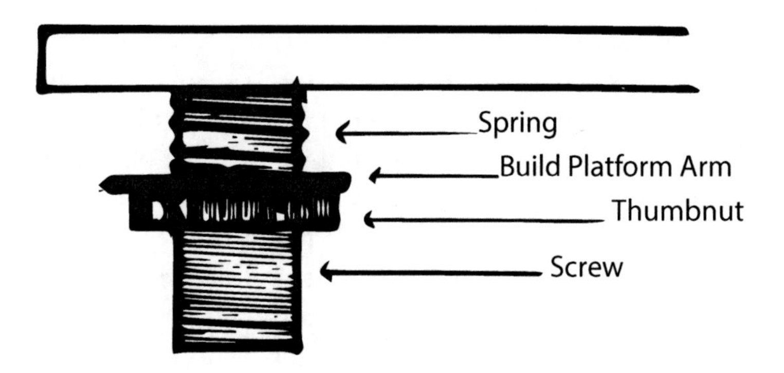

There are a number of ways to achieve a level bed. Machinists have traditionally used tram and dial gauges to achieve this with industrial equipment, and, in true trickle-down fashion, hobbyists have adopted these methods for their own use.

Taking another page from the machinist's handbook, some have invented and adopted a range of leveling techniques using feeler gauges. Newer methods such as interactive leveling procedures have also begun to emerge that are easier to learn and use on a regular basis.

Dial indicators work on a pressure gauge. When depressed, they read a number showing the distance of their depression. This means a dial indicator coasting evenly across a flat surface should read one, consistent value. There are a number of open-source dial indicator holders online that are available to be downloaded and printed. Once installed, allow the dial indicator to drag across various spots on the build platform and trim the adjustment screws so that it reads one consistent number across the bed.

Feeler gauges are another staple of the machinist's toolbox. They are precision-made metal shims of a specific thicknesses that can be used to measure, or "feel," the gap between two objects. Use a 1.0mm feeler gauge to level a build plate by putting it between the nozzle and plate and adjusting the height screws until there is a slight friction on the feeler gauge when slid back and forth.

Interactive leveling uses the printer's software to help calibrate the build platform. You can try this for yourself by downloading the interactive leveling file from our website (http://TheBookOn3DPrinting.com). As the object begins to print, slowly adjust each leveling screw under the build platform, watching as it goes. You want the object to print with a smooth, solid finish. If there are lines between the layers, the build plate is too low. If there are lines of light and dark overlapping plastic, or the edge looks wavy, then the build plate is too close.

Level

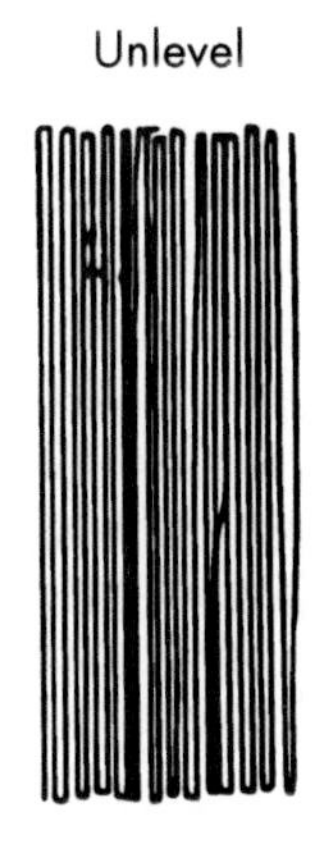

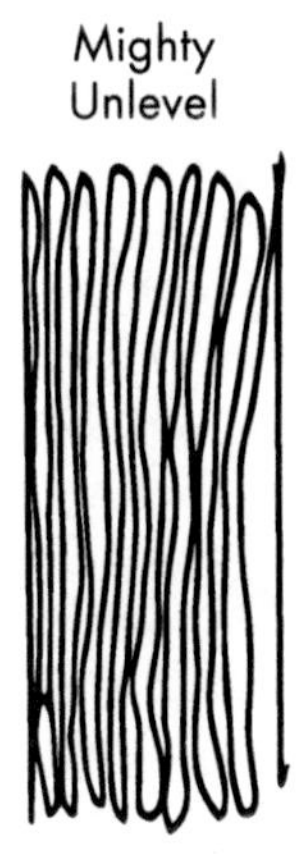

Thing removal

Depending on the material, size of your object, and surface treatment of the build plate, some things will be easier to remove than others. Ideally, objects will have enough adhesion to remain attached to the platform from the beginning to end of the print and gently pop off with light pressure when finished. However, some objects will require a little more effort. For those objects that don't come off easily, you will need to find a small, flat tool to act as a spatula. Gently slide the tool between the object and platform to pry it off.

There are a number of open-source printable tools on the web that accommodate standardized razor blade sizes. If you would like a prefabricated solution, the best option is one of the hobby spatulas used for vinyl decal removal.

10. The Extruder

The extruder is one of the more complex parts on a 3D printer. Exactly as its name implies, the extruder is the mechanism that pulls in the plastic filament, heats it up and squeezes, or extrudes, the hot plastic out a nozzle. The easiest way to understand this component is to follow the journey of the filament through the extruder.

Plastic filament is typically purchased by the spool, looking much like an oversized spool of thread made from the plastic wire used in weed whackers. The spool is usually attached to the printer on an axle so it can unwind as the plastic is used. Some printers come with spool holders integrated into the machine, while others are made to be free standing. Larger printers or spool holders can accommodate two or more spools for easy changing and storage.

Once again, let's use the analogy of the extruder as a hot glue gun and the spool of filament as the glue stick. Just like a hot glue gun melts a glue stick and extrudes glue at around 180° C, the extruder of a 3D printer pulls in solid plastic and squirts it out at 230° C. Unlike a hot glue gun, movement and flow control are not accomplished by moving our hand and squeezing a handle. In the case of a printer, these motions are controlled by a set of computer commands.

The extruder uses a clamping mechanism (called a "plunger") and a serrated wheel (called a "drive gear") to pull in the plastic filament. The plastic filament is fed between the plunger and the drive gear, which are pressed against each other to create friction on the filament. This allows the metal drive gear to bite into the soft plastic. From there, the plastic filament then begins its journey to the hot end. Depending on the extruder type, it could either be a quick ride or very long trip.

Parts of the Extruder

The extruder consists of two main parts: a *feeding mechanism* to carry the raw material into the machine and a *hot end* to melt and extrude it. Each of these assemblies has several internal components at work as well. A brief understanding of these parts will help with maintenance and troubleshooting of a 3D printer.

The *feeding mechanism* is the portion of the extruder that draws the printable material in and out of the machine. This section is made up of two components known as the *drive gear* and the *plunger*.

The *drive gear* is a metal wheel with serrated treading designed to grip and pull the plastic filament into the extruder. The treading can range from very fine and mild to very crude and aggressive. It is spun by a position-aware motor.

The *plunger* is the mechanism that keeps the plastic filament pressed against the serrations on the drive gear, allowing the extruder to keep a firm hold on the plastic. The plunger consists of an arm to push the filament against the gear, a spring or other source of pressure to keep the arm pressed against the filament, and often a set screw or spring release used to adjust the pressure against the filament.

The *hot end* refers to the parts of the extruder that

do the melting and squirting of the plastic filament. This section consists of a heating element, a temperature sensor, a nozzle, and a block to keep it all together.

The *heating element* is pretty straightforward in its purpose. It is a small electronic component capable of producing high temperatures. Heating elements are usually integrated into extruders in one of two ways. The first method is a cartridge-based heating element, which looks like a half-inch metal straw that slides into a predrilled hole in the hot-end. The second method involves wrapping wire around the bulk of the extruder's metal tubing and covering it in a non-conductive, heat-resistant medium like ceramic adhesive.

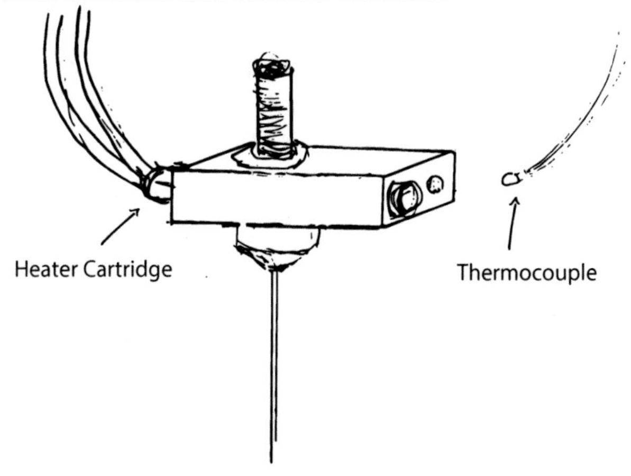

Temperature inside the hot end is monitored using a *temperature sensor*. There are two types of temperature sensors widely used in 3D printers, they are *thermistors* and *thermocouples*. Both sensors work by changing the flow of electricity based on temperature. The computer then reads the flow and can compute the accurate temperature. Thermistors contain a semiconductor which changes resistance when heated; thermocouples change voltage using a junction of dissimilar metals. Typically thermocouples have a wider range where thermistors have higher sensitivity to small changes.

The *heater block* is a heat-resistant piece of aluminum block or high-heat plastic (such as PEEK) to which the nozzle, feed barrel, heating element, and temperature sensor all connect. It acts as both an insulator for the nozzle and as a heat dispersion element, which protects the other components around it. Running through the heater block is the *barrel*, which is a threaded metal tube that serves as a heating chamber for the plastic on its way to the nozzle. It is in the barrel that plastic filament reaches its *glass transition point*, which is just a fancy way of saying the temperature at which it liquefies. From there, the pressure of the plastic behind it pushes the liquid out the nozzle.

There are a number of standardized nozzle sizes, such as 0.60mm, 0.50mm, 0.40mm, 0.35mm, 0.30mm, and 0.25mm. The smaller the nozzle, the more accurately fine details can be printed. However, a bigger nozzle means objects can be printed faster but with less detail. The most commonly used sizes are .5mm and .4mm, as they offer the best balance of speed and detail.

Most extruders feature a fan somewhere in the hot end assembly. Fans helps to reduce residual heat from the heating element. This keeps all parts functioning at their optimal temperatures and prevents heat-degradation of the object being made.

Extruder Types and Designs

There are two popular extruder design for 3D printers: *direct drive* and *Bowden*. The former performs more like a front-wheel drive vehicle, where the weight of the car body is being pulled by the engine. The later works more like a rear-wheel drive vehicle, where the engine pushes the weight of the car body. Another way to think about this is that a direct-drive extruder *pulls* the filament into the hot end, where a Bowden extruder *pushes* filament into the hot end.

In a *direct-drive* configuration, the feeding mechanism is positioned immediately above or directly over the hot end. The plastic is fed into the top travels through the feeding mechanism, and then makes a very short jump (a few centimeters or so) into the hot end. This configuration is currently the most widely adopted design. This is partially because it is an older design, and also because direct-drive designs have fewer parts. This makes them easier to handle from a maintenance perspective. One downside to the direct-drive method is that a motor must be mounted on an axis, adding weight to the gantry. This limits the printing speed of the machine.

Bowden extruders place the feeding mechanism away from the hot end. They are connected by a slippery plastic tube that is secured at both ends. The filament slides through the feeding mechanism into the tube. Using the tube as a guide, it travels anywhere from a few inches to a foot until entering the hot end. Bowden type extruders can run at faster speeds because they move the heavy motor off of the moving gantry, leaving only the lightweight hot end. However, the Bowden design is less documented and by nature requires more parts. For this reason, Bowden-style designs may be best for more experienced users.

Spring vs Plunger Designs

In order for the drive gear teeth to grip the filament, there needs to be a force pushing firmly against the filament. Two ways of accomplishing this are *spring-loaded* and *plunger-based* extruder designs are two ways of accomplishing this.

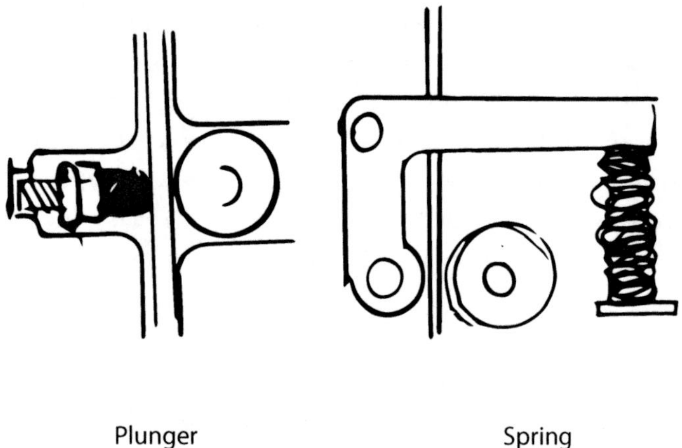

Plunger Spring

A spring-loaded extruder uses a spring and arm to push the filament against the drive gear. Typically there is a bearing at the end of the arm to help keep the filament moving against the drive gear. With too much pressure, the drive gear can strip the filament away to the point where it is too thin to be gripped. This will result in an extruder jam. The bearing works to prevent this.

A plunger design works by pressing a hard rod, called a plunger, against the filament. A rubber elastomer and set screw are used to set the pressure against the filament. This design works more reliably with softer plastic like ABS. The mechanism has a tendency to break with PLA and harder materials. This is because the mechanism has little room to accommodate the stiffer filament.

Retraction and Residual Heat

One of the inherent issues with a nozzle full of liquified and pressurized plastic is keeping ooze under control. This is when excess plastic seeps out of the nozzle, sometimes causing unwanted blemishes and inconsistencies. Most commonly, plastic ooze will leave tiny dots, called pimples, all over the object. This can ruin the smooth finish of the surface.

To help combat ooze, a function called *Retraction* is built into the software of a 3D printers. This feature performs exactly as it sounds, retracting the filament and pulling it slightly back when moving the print head around. This removes some of the pressure on the liquified plastic and prevents it from unintentionally oozing out of the nozzle.

With direct drive extruders, placement of the feeding mechanism directly over the heating element can lead to a buildup of residual heat. This can cause issues with retraction. As the drive gear pulls the filament back into the extruder, the extended heat soak can cause filament to soften too quickly. Soft filament will stretch rather than retract, resulting in an extruder jam.

Filament cleaner

There are a number of excellent filament suppliers, most of which offer high-quality filament for 3D printing. However, there is the chance of occasionally receiving a spool of filament that has inconsistencies or defects. One of the issues you may run into with filament is not with the quality of the plastic, but with dust buildup.

The average nozzle opening on a 3D printer hot end is less than half a millimeter wide. This means that even the slightest speck of dust can cause a major clog in the system. Spools and printers will collect dust in most environments, but if your printer is in your workshop or stored somewhere that dust is likely to accumulate, then a filament cleaner may be a good idea.

A filament cleaner is a small container with either a piece of sponge or cloth inside that the filament will pass through on its way to the extruder. The sponge or cloth will wipe down the filament as it is pulled in, keeping dust particles out of your extruder and easily cleaned up. The filter inside the filament cleaner should be changed after every hundred hours of printing.

11. The Electronics

Like many digital devices, a 3D printer uses a microprocessor to control its various electronic components as it follows the steps required to print CAD files. However, 3D printers are a very different animal from laptops or computers that require voltage, circuitry and electronic componentry powerful enough to compute and process the mathematics involved in three-dimensional design.

A basic understanding of the electronic components of a 3D printer will help you address any electrical bugs and performance issues originating with these parts of your machine.

As in many electronic devices, the *motherboard* houses the "brains" of the 3D printer. This is where the instructions that control the printer are processed. Much like our own, human brain regulates the complex functions of our bodily systems; the electronics on the motherboard tell the motors to spin, the heater cartridge to warm up, and the extruder to release melted filament.

Commands from the software are executed by firing electrical signals from the *Integrated Circuit* (IC). This is the microprocessor component, where the machine's instructions are calculated and interpreted. The microprocessor then tells the various components what to do and when to do it, all while listening for feedback. Constant awareness of temperature and carriage position, monitored through sensors inside the printer, allow the printer to remain aware of the location of it's physical parts during the program.

You can imagine this process of constant feedback to be much like our own abilities as organic machines. For example, if you were to run a bath for yourself, you might initially turn on the faucet by approximating a setting somewhere between cold and hot. This is based on where you think the desired temperature will be found. Next, you might check the temperature with your hand, feeling and deciding like Goldilocks if the water is too hot, too cold or just right. If the bath is too hot or too cold, you can adjust the faucet accordingly and check again. This process will be repeated until a comfortable temperature is found and continued throughout the bath as more hot water is added to keep the tub at the ideal temperature.

In the case of a 3D printer, the electronic components of the motherboard manage the heating process very similarly. A command from the software tells the heater to warm up (like turning on the faucet) and the temperature sensor provides feedback to let the printer and software know how warm the hot end has gotten. When it reaches a temperature that is just right (as defined by the software or digital file being printed), it will maintain that level by running just enough electricity through the component to keep the temperature constant.

Foundation: RAMPS Board

If you are thinking about building your own printer, or if you have a kit-type printer, it is helpful to know a bit about the board inside your machine. Motherboards for 3D printers are as varied as the machines they control. They range from open-source community products, like

Arduino shields, to proprietary controller boards from Makerbot or 3D Systems.

The RepRap Arduino MEGA Polulu Shield or *RAMPS* board was born of the RepRap project, as its name suggests, and is an open-source shield that easily fits on an Arduino Mega board. RAMPS was designed with the goal of building a low-cost, small package containing all the electronic components necessary to control a 3D printer. Needless to say, that goal was achieved. RAMPS continues to be one of the most popular choices of motherboard for homebrewed machines, and for a number of printers on the market. Some of the proprietary boards now used inside Makerbot and Ultimaker began as derivatives of the open-source RAMPS board.

It is somewhat uncommon for the electronics of a 3D printer to require maintenance, unless an external event like a power surge has affected the machine. It is far more common that the components and sensors controlled by the motherboard will need to be adjusted or replaced. However, on rare occasions, a board can burn out and will need to be replaced. Familiarity with these parts will make these tasks easier and give you a better understanding of the internal workings of the machine.

Electronic Components

End stops are tiny switches that, when pressed, complete a circuit. This feedback is used to tell the motherboard that the carriage is at its home position, or at the origin point of the axes. Before starting a print job or moving its parts, a printer will "home" its components by moving them back to this starting position. The home position is defined by the placement of three end stops, one for each axis (X, Y and Z). When the carriage reaches the home position, it depresses the switch, telling the controller board the locations of the moving parts.

Once the printer is aware of its position, it can begin issuing commands to move the print carriage, build platform, and other components. Precise motors that can be articulated by a specific number of units, or *steps*, by a digital controller. This is how the printer is able to guide its extremities to the points necessary to build the object.

A *stepper motor* is used to execute these movements. Stepper motors contain a magnetically polarized spindle inside a charged coil. The coil issues electric pulses causing the central spindle component to articulate one step at a time. From a software perspective, this makes it possible to calculate the exact number of steps required to reach the home position or any other point on the build platform.

To move carriages, platforms and print heads, stepper motors need a good deal of power. This can put a lot of strain on a circuit board. To solve this, a tiny circuit board called a *stepper motor driver* is placed on the motherboard. These help regulate the power drawn by the stepper motor. If a stepper motor gets strained beyond its capacity, or gets pushed beyond its performance parameters, the stepper motor driver will blow before more serious (and expensive) damage is done to the motherboard.

The motherboard also receives feedback from *temperature sensors*. Two of the most popular temperature sensors used in 3D Printers are thermistors and thermocouples. These sensors work by changing the flow of electricity based on temperature. The controller then reads the electric flow and computes the temperature. This is how printer software is are able to precisely regulate the temperature of the hot end, build platform or other heating element.

Thermistors use a semiconductor to change electric

current resistance. This means they resist, or choke off the flow of electricity as the temperature gets higher. *Thermocouples* work somewhat differently, changing voltage in response temperature by using a junction of dissimilar metals. Thermocouples have a wider temperature sensing range compared to thermistors. However, thermistors have a higher sensitivity to small changes, which can be an advantage for a precise machine like a 3D printer.

A *power supply* is used to connect the printer to an external power source and supply electrical current to all of the processes happening inside a 3D printer. A power source unit is provided with a printer. If building your own machine, it is important to calculate the electrical load requirements of your internal components and to purchase an appropriately powerful unit. This information is usually available from the component supplier or manufacturer.

Replacing a Thermosensor

If there is one part you will need to replace on your extruder, it's the temperature sensor. The most popular types of temperature sensors often contain glass beads that are very prone to breaking. Installing a new thermosensor on the circuit board itself is not a difficult thing to do. This usually involves plugging new wiring into the headers. The challenge lies in properly securing the sensor to the hot end.

If the thermosensor comes loose it will not read the temperature correctly. This can cause the printer software to think the heater is not hot enough and drive more energy to the heating element. The hot end will get hotter and hotter while the sensor reads the air temperature. An incorrectly installed thermosensor runs the risk of getting too hot and melting the critical components around the hot end and heating block. This poses a fire hazard and may critically damage the machine.

To properly secure a thermistor or thermocouple, use a strain relief on the hot end so that any tugging or yanking that happens as the carriage moves around does not pull the sensor off the end. Securely attach the sensor under several layers of *Kapton tape*, which is capable of sustaining the high temperatures of the hot end.

Some temperature sensors come embedded inside a screw which fits into the heating block on the hot end. We highly recommend this type of sensor for its superior durability. The sensitive parts on these sensors are safely embedded inside by screwing the sensor into the heating block, offering greater reliability.

Epilogue

The days of dreaming about instant objects may soon be over.

What we now know as 3D printing represents only the beginning of a world where the objects we imagine can become realities we can hold in our hands. This idea was once the stuff of science fiction, but the research and hard work of creating machines that can make *anything* we desire is well underway. Much like the lightbulb, the automobile and the Internet, the incredible inventions that await us will shape our way of life and forever change the way we think about our world.

Imagine for a moment the vast benefits of on-demand things. They will be far-reaching for sure, giving us the ability to manipulate materials that are conductive, pliable, wearable and even edible. These are not possibilities, but eventualities, as even now, scientists and hobbyists around the world are working to develop ways of extruding, sintering and otherwise forming these things into yet unimagined creations. The results will touch us all, as we are soon able to digitize, customize and commoditize the physical realm as we have done with the whole of human knowledge now available instantly online.

Consider the impact of just a few of these breakthroughs, like the instant assembly of food. It is now possible to program machines to build pizzas, cakes and confections from sets of computer-controlled instructions. In this reality we find the first steps toward future devices equipped with raw ingredients, heating elements and even natural and chemical substances that will make these same things, and much more, from scratch according to digital recipes. With such devices in our homes, the microwave oven would appear to be a relic of a primitive past as we conjure up gourmet meals, family favorites, and everyday staples with a few simple commands.

Similarly, personal items like clothing may be produced from flexible, fabric-like materials that will be crafted into the fashions we may soon be wearing. This future will be a giant leap forward for individuality and personal expression when compared to the mass-produced and poorly-made clothes found in today's department stores. This will be a dawn of new possibilities for consumers, but also designers and retailers, who will make endless styles available, freed from the costly burden of manufacturing clothes in large volumes. Instead, digital fashions will be matched to our physical shapes, customized to our personal tastes (a broader shoulder here, a longer hem there), and spun from our own machines as a one-of-a-kind wardrobe.

As you read this, perhaps on a tablet or e-reader, you may have momentarily forgotten that your electronic device is rather quickly heading toward obsolescence. This is the nature consumer electronics, which are refreshed and reinvented at a dizzying pace. This progress will not slow down in the future that awaits us, but the ability to fabricate electronics in our homes will eventually ease the burden we all bear to keep up with the latest trends. Imagine the ability to download not only software updates, but also hardware upgrades. The ability to make componentry that can be installed at home by the product owner would result in a far lower cost of ownership for these products we hold dear. Machines able to work in multiple materials, molding plastics as well as circuitry, could one day distribute the manufacturing of electronics from factories far away to desktops everywhere.

Perhaps some of the most exciting achievements of this future world will be in the medical field. Today, scientists have taken the initial steps to apply 3D printing techniques to create human tissue for transplant. Using real and simulated organic materials, doctors will one day replace lost or damaged body parts ranging from ears and joints to internal organs. This will change the nature of medical care and open the possibility for new industries focused on healing and even improving the human body. It may one day be possible to design our own physical enhancements from the comfort of our living rooms.

These things, and many more, are the logical next steps in the evolution of additive manufacturing. Products, services and realities once impossible will become commonplace in the same way the highways, airports and instant communications we now enjoy were once nothing more than fantasies. This is why we wrote *The Book on 3D Printing*, not merely to explain the details of today's 3D printers, but to introduce this technology to the many people, like you, who will embrace its potential and advance its capabilities.

No matter the reason for you interest in 3D Printing, we hope you have enjoyed learning more about what it can do. Perhaps we have even been successful in inspiring you to explore these ideas further in your own work or hobbies, or perhaps not. Either way, we hope this book has been helpful in your pursuit. Along your journey, do stop in at TheBookOn3DPrinting.com for additional resources and ideas you may find helpful.

Here's to a future filled with wonders!

Let's make it together.

Glossary

of 3D Printing Terminology

Acrylonitrile Butadiene Styrene (ABS): A plastic commonly used as filament for 3D printers; requires a heated surface for proper printing

ABS glue: An adhesive solution made by dissolving ABS plastic in acetone; used by hobbyists for gluing parts together and improved build platform adhesion

Additive Manufacturing: The process of building a three-dimensional object by forming a material into layers

Axis (or axes): A mechanical assembly that allows linear movement or mechanical components

Bed: Also known as the build plate

Belt (or timing belt): A cogged band or loop, usually made of reinforced rubber, that facilitates the movement of the carriages along the X and Y axes.

Belt-driven: Movement accomplished by belts and pulleys; faster than a screw driven but with less precise positioning

Build platform: The surface on which a 3D printer will form an object; sometimes called a build platform

Build volume: The total available space inside a 3D printer for forming objects; measured in three-dimensions

Cartesian Coordinate System: The method used by 3D printers to identify points in space along three directional planes; points are plotted from an origin or (0,0,0) [x,y,z]

Computer-aided Design (CAD): The use of software to assist in the design or modification of a three-dimensional object

Computer Aided Manufacturing: The use of software to control the operations of machine tools in the creation of physical goods

Carriage: A moving assembly that travels on one of the axes in a 3D printer; components are attached to a carriage to provide motion

Delamination: When two or more layers of an object lose adhesion and come apart

Drive gear: A gear located inside the extruder used to pull filament into the print head

End stop: A mechanical switch used to sense the home location of the printer carriages

Extruder: The part of printer that pulls in, heats up

and spits out the plastic filament

Extrusion: The process of forcing material through a small opening to form a shape

Facet: A small plane surface, hundreds or thousands of which make up the surface of a digital object; can be compared to the surface or "cuts" in a diamond

Filament: A thread-like plastic on large spools used as raw material by FFF 3D printers

Fused Deposition Modeling (FDM): A patented additive manufacturing process found in most consumer 3D printers; involves the computer numerically controlled extrusion of thermoplastic filament onto a build platform

Fused Filament Fabrication (FFF): The legally unconstrained name for Fused Deposition Modeling; originally adopted by the RepRap community

G-code: The programming language used to give instructions to numerically controlled machine tools, like 3D printers

Gantry: The system for mounting the axes, carriages and linear motion components inside a machine

Hot end: The part of the extruder that heats filament to its melting point; consists of a heating element, nozzle, and filament tube

Infill: The variable that defines the density of the internal support structure of FFF printed objects; the higher the percentage of infill, the denser the object

Kapton tape: A polyimide adhesive film by DuPont which remains stable at very high temperatures; used on heated build platforms to aid adhesion, as well as for extruder insulation

Layer height: Also known as object resolution, refers to the height of the horizontally printed layers of an additively manufactured object, typically measured in microns

Micron: A unit of measure equal to one millionth of a meter

Object resolution: see layer height

Parametric modeling: a method of creating 3D or CAD designs using a scripting language to define the features of geometric solids

Polyaryletherketone (PEEK): A thermoplastic polymer that is resistant to high temperatures

Polylactic Acid (PLA): A thermoplastic polyester made from starch, such as corn or other crops; a popular FFF filament.

Post processing: Any work done to an object after

the machine has finished printing; can include sanding, painting, drilling, and bathing

Raft: Material printed alongside an object to impede warping; objects are built on top of a raft, or mesh netting, made from disposable material

RepRap: A self-replicating rapid prototyper, or a machine capable of reproducing it's own components; also the name of the RepRap Project which produced versions of the technology now found in many 3D printers

Screw-driven: movement accomplished by moving a nut along a threaded rod; slower than belt driven systems, but capable of more precise positioning

Slicer: A software program that converts CAD files into G-code instructions

Slicing: The process of converting a 3D model into instructions for a computer numerically controlled (CNC) device to follow

Stepper motor: An electric motor that can produce precise partial rotations

Stereolithography (SLA): a method of 3D printing that produces objects by forming object layers from a material that is cured by ultraviolet light

Selective Laser Sintering (SLS): A method of 3D

printing that produces objects by using a laser to forming layers from a liquid or powder

STL file: The file format used by many 3D printers; describes the surface geometry of an object in binary code

Thermistor: A type of resistor used in 3D printers to provide temperature feedback in a 3D printer; thermistors are read by measuring the change in current at varying temperatures

Thermocouple: A sensor consisting of two conductors that generate a current when heated; thermocouples provide temperature feedback by measuring the voltage produced at varying temperatures

Thermoplastic: A type of plastic, or polymer, that becomes pliable at a specific temperature

REFERENCE CHART | Object Resolution & Layer Height Range

Object Resolution	Layer Height (microns)	Layer Height (mm)
High	25 - 150	0.025mm-0.15mm
Medium	160-250	0.16mm-0.25mm
Low	260-340	0.26mm-0.34mm

*A sheet of paper is about 100 microns, or 0.1 millimeter thick.

Glossary